EDU

EXPLORING MATHEMATICS ON YOUR OWN

Permutations and Groups

EXPLORING MATHEMATICS ON YOUR OWN

Permutations and Groups

C.D.H. COOPER

JOHN MURRAY · LONDON

Typeset by Tecprint Ltd, Loughborough, and
printed by offset in Great Britain
by Martin's Printing Works, Berwick-on-Tweed

0 7195 2909 3

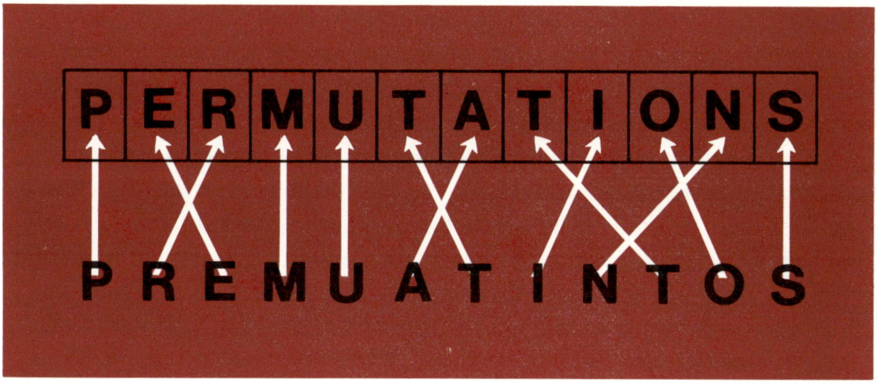

Permutations

Arrangements

The order in which things are arranged is usually very important. If the pages of this book had been put together in the wrong order, you would not have been able to read it properly. Oredr is of graet ipmortacne in the wya wodrs are put togehter: even if we only vary the order of the letters a little, the effect is quite strange and borders on the unintelligible. The lists of symbols 'aaaaa c dd eeeee gg hhhh iiiii ll m nnnn oo p rrrrrr sss Tttt uu v w yy' conveys no meaning whatsoever, yet it consists simply of the first sentence of this paragraph rearranged (in this case into alphabetical order).

Even a fairly small number of objects can be put into many different arrangements. For example, there are 24 different ways of arranging the digits 1, 2, 3 and 4. To put this another way, we can make up 24 different four digit numbers using each digit once.

1234	2134	3124	4123
1243	2143	3142	4132
1324	2314	3214	4213
1342	2341	3241	4231
1423	2413	3412	4312
1432	2431	3421	4321

If you think about it, you will see that the number of arrangements of a collection of different objects does not depend on what the objects are. It depends only on the number of objects. Take any four different things and you will find that you can arrange them in exactly 24 ways.

It is possible to work out the number of arrangements of a collection of objects without having to write down the arrangements themselves. In the above example we have four digits to arrange. There are thus four digits that can be put into the first place in an arrangement. These correspond to the four columns in the above table. Now having chosen the first digit we will be left with three digits to choose from in filling the second place. We therefore have three ways of filling up the second place *for each of the four ways of filling up the first place.* There are thus $4 \times 3 = 12$ ways of filling the first two places. There is very little choice in completing the arrangement. When we come to put in the third digit we have only two left to choose from and so there are two possible choices for the third place for each way of filling up the first two places. There are thus $4 \times 3 \times 2 = 24$ ways of filling the first three places. Finally, there is no choice (strictly speaking there is one choice, or one alternative to 'choose' from) in filling the fourth place, because with three out of the four digits having been chosen to go in the first three places, there is only one digit left and it must go last.

Given any four objects, therefore, there are $4 \times 3 \times 2 \times 1 = 24$ different arrangements of them. This argument extends to any number of things. The general formula is $n(n-1)(n-2) \ldots \times 3 \times 2 \times 1$ for the total number of arrangements of n things. For example, the five objects *, △, +, ○, □ can be arranged in $5 \times 4 \times 3 \times 2 \times 1 = 120$ ways.

There is a shorthand way of writing down $n(n-1)(n-2) \ldots \times 3 \times 2 \times 1$. We denote it simply by $n!$ In words, we call this *factorial n*. It has been said that the reason for this exclamation mark is that $n!$ is usually a very big number. Whether or not this explanation is correct, it certainly is true that as n increases in size, $n!$ gets very big indeed.

Table for factorial *n*

n	$n!$
1	1
2	2
3	6
4	24
5	120
6	720
7	5 040
8	40 320
9	362 880
10	3 628 800

From the table we can see, for example, that the digits 0, 1, 2, . . . , 9 can be arranged in over 3½ million ways. The number of ways of arranging a pack of 52 cards is 52! which is an extremely large number – greater than 8×10^{67}.

Permutations

Shuffling a pack of cards is an operation which changes one arrangement of the cards into another. Operations of this type are called *permutations.* For example the operation of taking the top card from a pack of cards and putting it on the bottom is a permutation. The effect of this permutation on one particular arrangement is illustrated below, where the top row illustrates the original arrangement and the bottom row illustrates the result of applying, or carrying out, the permutation.

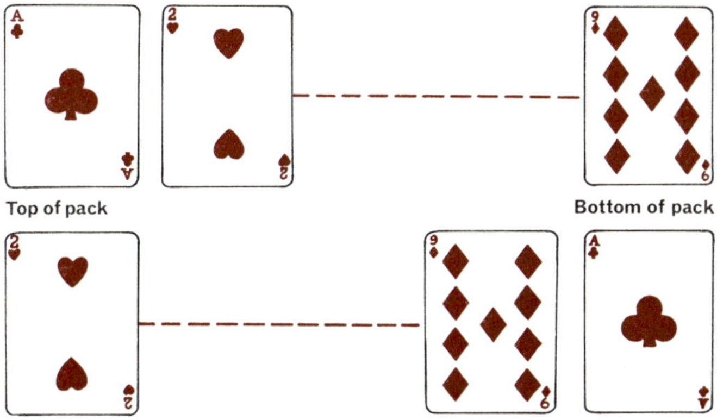

There are, of course, far more complicated permutations than this. However their complicated nature arises more from the difficulty of expressing the rearrangement in ordinary language than from the permutation itself. There is, fortunately, a concise and easy to follow symbolic notation for permutations. By expressing permutations in this notation, one can manipulate them in a way that would be impossible if they were described in words.

To see how this notation works, consider the following rearrangement. The original arrangement appears first and the new arrangement is depicted underneath.

*	△	+	○	□
△	□	*	○	+

In words, we could express this permutation by saying:

*	is replaced by	△
△	is replaced by	□
+	is replaced by	*
○	is replaced by	○
□	is replaced by	+

A more concise way of expressing this would be to use an arrow to represent the phrase 'is replaced by':

* → △
△ → □
+ → *
○ → ○
□ → +

Fig. 1

A further refinement is to combine the above five symbolic statements into one diagram. The symbol '→' still represents the phrase 'is replaced by':

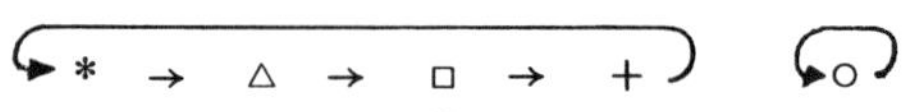

Fig. 2

Not only does Fig. 2 take less space to write down, it makes the information it contains more readily available. It can be seen at a glance that if the permutation is applied four times, each object is replaced by itself (follow a sequence of four arrows from any object and you return to that object). This information is not quite so immediately apparent from Fig. 1. The following three permutations, depicted in the manner of Fig. 1, look rather similar.

(i)	(ii)	(iii)
$A \to E$	$A \to E$	$A \to E$
$B \to D$	$B \to D$	$B \to A$
$C \to F$	$C \to F$	$C \to F$
$D \to C$	$D \to B$	$D \to D$
$E \to B$	$E \to A$	$E \to B$
$F \to A$	$F \to C$	$F \to C$

Fig. 3

However when depicted in the manner of Fig. 2, they can be seen to be quite different.

(i) $A \to E \to B \to D \to C \to F$

(ii) $A \to E$ $B \to D$ $C \to F$

Fig. 4 (iii) $A \to E \to B$ $C \to F$ D

We can see immediately from Fig. 4 that the permutation (iii) *fixes* (i.e. replaces by itself) one object while the other two do not fix any object. Furthermore we can see from Fig. 4 that if the second permutation is carried out twice, every object returns to its own place while this is not the case with permutations (i) or (iii). All of this information is present in Fig. 3, but it lies beneath the surface.

Cycle notation

The system of notation used in Fig. 4 can be called a *cycle notation* because it breaks up the set of objects into cycles. Permutation (i) above consists of a single cycle while permutations (ii) and (iii) each involve three cycles. What is usually termed *the cycle notation* is a slight refinement of that used in Fig. 4.

For a start, the arrows can be removed from the diagram, provided that we separate the cycles from one another. The above three permutations can be represented as follows:

(i) $(AEBDCF)$; (ii) $(AE)\,(BD)\,(CF)$; (iii) $(AEB)\,(CF)\,(D)$

We adopt the convention that each object is replaced by the one on its right-hand side, except for the last object in each cycle which is replaced by the first. This convention removes the need for arrows.

A second refinement is that we remove any cycles which contain just one object. We adopt the convention that any object which does not appear in the cycle notation for some permutation, is fixed by that permutation. Permutation (iii) above, can be written simply as $(AEB)\,(CF)$.

Examples. Write the following permutations in cycle notation:

(a) $1 \to 3$, $2 \to 6$, $3 \to 4$, $4 \to 2$, $5 \to 5$, $6 \to 1$

Answer: $(13426)\,(5) = (13426)$

(b) 1→5, 2→6, 3→2, 4→1, 5→4, 6→3
Answer: (154) (263)

(c) 1→6, 2→2, 3→4, 4→1, 5→5, 6→3
Answer: (1634) (2) (5) = (1634)

(d) 1→5, 2→2, 3→3, 4→4, 5→1, 6→6
Answer: (15) (2) (3) (4) (6) = (15)

(e) 1→1, 2→2, 3→3, 4→4, 5→5, 6→6
Answer: (1)(2)(3)(4)(5)(6) = ?

This last example raises one small difficulty. If we adopt the convention that we can remove any cycle which contains just one object, we may be left with nothing at all. This would be the case with the so-called *identity permutation* which replaced every object by itself. To overcome this difficulty we use the special symbol 'I' to represent the identity permutation. Thus the answer to example (e) is simple 'I'.

EXERCISE SET 1

1. Write the following permutations in cycle notation:
   (a) 1→3, 2→4, 3→5, 4→1, 5→2
   (b) 1→4, 2→5, 3→3, 4→2, 5→1
   (c) 1→2, 2→4, 3→5, 4→1, 5→3
   (d) 1→1, 2→5, 3→3, 4→4, 5→2
   (e) 1→1, 2→2, 3→3, 4→4, 5→5
   (f) 1→2, 2→1, 3→3, 4→5, 5→4

2. Which of the following is not a permutation?
   (a) 1→3, 2→4, 3→2, 4→1
   (b) 1→1, 2→4, 3→3, 4→1
   (c) 1→3, 2→2, 3→1, 4→4

3. Which of the following permutations is different from the other two?
   (a) (1428) (375)
   (b) (1824) (573)
   (c) (753) (2814)

Uniqueness rules

According to the rules we have set down for writing a permutation in cycle notation, there are many ways of writing down the same permutation. For example, (1234), (2341), (3412) and (4123) all represent the permutation $1 \to 2$, $2 \to 3$, $3 \to 4$, $4 \to 1$. It is desirable to have a unique way of writing down a permutation so that we can tell at a glance when two permutations are equal. We can achieve this when the objects are the numbers 1,2,3,... (as they will be for the remainder of this book) by adopting the convention that the smallest symbol in each cycle is to be put first. Thus of the four alternatives above we choose (1234) to represent the permutation.

This does not completely settle the uniqueness problem however, since the permutation

$$1 \to 4, \quad 2 \to 3, \quad 3 \to 2, \quad 4 \to 1$$

could be represented by (14)(23) or (23)(14). In both cases the smallest number in each cycle is put first. We choose one of these and reject the other by adopting the convention that the various cycles in a permutation are ordered according to their first symbols. The cycle which has the smallest first symbol comes first, and so on. This means, for instance, that the symbol 1 will always appear first, if it appears at all.

Example. $1 \to 1$, $2 \to 8$, $3 \to 2$, $4 \to 3$, $5 \to 9$, $6 \to 7$, $7 \to 6$, $8 \to 4$, $9 \to 5$ when written in cycle notation is (2843)(59)(67).

Summary of the cycle notation system for representing a permutation on 1,2,3, . . . , n

General form of a permutation:

$$(a_1, a_2, \ldots, a_r)(b_1, b_2, \ldots, b_s) \ldots$$

Meaning: $a_1 \to a_2$, $a_2 \to a_3$, . . ., $a_r \to a_1$

$b_1 \to b_2$, $b_2 \to b_3$, . . ., $b_s \to b_1$

. . . .

i.e. each symbol is replaced by the next symbol to the right, except for the last symbol in each cycle which is replaced by the first symbol in the cycle.

Symbols which do not appear are fixed.

Uniqueness conditions: a_1, b_1, . . . are the smallest numbers in their respective cycles. a_1, b_1, . . . are in ascending order.

Exception to the above: 'I' denotes the identity permutation which fixes everything.

EXERCISE SET 2

1. Which of the following violate the uniqueness conditions for cycle notation?
   (a) (2875)(364)(19)
   (b) (1987)(456)
   (c) (17925)(4836)
   (d) (15)(26)(37)(89)

2. Write down, in cycle notation, all of the permutations on the objects 1,2,3. (If you write them down correctly, observing the uniqueness conditions, there will be no danger of repetitions.)

The number of permutations

If you had correctly answered question 2 in the above set of exercises you would have discovered that there are six permutations on three objects. Remember that this is the number of different arrangements on three objects. The reason for this is that we can, if we wish, regard an arrangement as a permutation – as the permutation which produces that arrangement, having started with the arrangement 1,2,3. Thus the arrangement 2,1,3 can be identified with the permutation $1 \to 2$, $2 \to 1$, $3 \to 3$ which in cycle notation becomes (12). Conversely, every permutation can be thought of as an arrangement – as the arrangement that results when the permutation is applied to the arrangement 1,2,3. Arrangements and permutations can be paired off with one another in this manner and so there must be the same number of each – in this case, six.

Obviously the above argument holds no matter how many objects there are. We may conclude, therefore, that the number of permutations on n objects is equal to the number of arrangements, viz. $n!$

The set, or collection, of all permutations on the objects $1,2,3,\ldots,n$ is denoted by S_n. Thus S_n contains $n!$ permutations. For example, S_4 contains $4! = 24$ permutations: there are 24 permutations on the objects 1,2,3,4.

S_4 = set of permutations on 1,2,3,4

I	(12)	(13)	(14)	(123)	(132)
(12)(34)	(23)	(24)	(34)	(124)	(142)
(13)(24)	(1234)	(1324)	(1423)	(134)	(143)
(14)(23)	(1243)	(1342)	(1432)	(234)	(243)

The 24 permutations on 1,2,3 and 4 are set out in the above table. Among these 24 permutations there are five distinct types of permutation (corresponding to the way in which they are grouped in the table). There is the very special and unique identity permutation, I, six permutations of the form (xx), eight permutations of the form (xxx), six permutations of the form $(xxxx)$ and three permutations which are made up of two cycles each, of the form $(xx)(xx)$.

The pattern of the cycles of permutations can provide a great deal of information about the permutations themselves, as we shall see in later chapters. This pattern is called the *cycle structure* of the permutation, or more simply the *shape* of the permutation. The shape of a permutation can be depicted by replacing each of the symbols 1,2,3, . . . in the cycle notation by the symbol 'x' and rearranging the cycles in order of size. (The shape of the identity permutation, I, is simply represented by 'I'.)

For example, the permutations (12)(365), (16)(253), (143)(25), (26)(435) and (264)(35) all have the same shape, viz. $(xx)(xxx)$.

EXERCISE SET 3

1. Write down the cycle structure, or shape, of the following permutations.
   (a) (14)(25)(37)(698)
   (b) (1437)(25)(89)
   (c) I
   (d) (193)(27)(4856)

2. Write down all of the permutations in S_5 (i.e. on the objects 1,2,3,4,5) which have the following shapes.
   (a) (xx)
   (b) $(xx)(xx)$
   (c) $(xx)(xxx)$

3. Write down all of the shapes that occur among permutations in S_6.

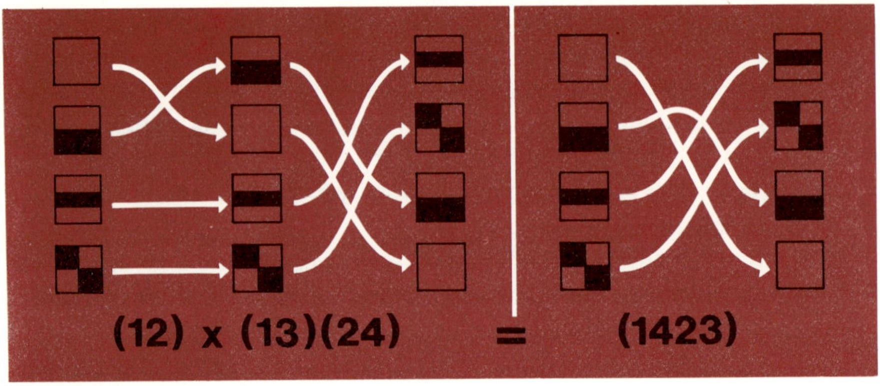

Multiplying Permutations

Following one permutation by another

If one permutation is carried out after another, the total effect will be equivalent to a single permutation. So if we first perform the permutation

$$(124) = 1 \to 2, \quad 2 \to 4, \quad 3 \to 3, \quad 4 \to 1$$

and follow it with the permutation

$$(13)(24) = 1 \to 3, \quad 2 \to 4, \quad 3 \to 1, \quad 4 \to 2$$

the symbol 1 will be replaced by 2 by the first permutation, and this 2 will then be replaced by 4 by the second permutation. We can write this as $1 \to 2 \to 4$. Investigating the movements of all four symbols we find that

$$1 \to 2 \to 4, \quad 2 \to 4 \to 2, \quad 3 \to 3 \to 1, \quad 4 \to 1 \to 3.$$

If we are interested only in the final effect and not in the way it comes about, we can write

$$1 \to 4, \quad 2 \to 2, \quad 3 \to 1, \quad 4 \to 3$$

which in cycle notation is (143).

Thus, as far as the final result is concerned, the effect of carrying out the above two permutations in succession is the same as if we had performed the single permutation (143).

The product of two permutations

The operation of carrying out one permutation after another is called *multiplying* the permutations, and the permutation that results from this is called the *product* of the two permutations. For example we have just seen that the product of (124) and (13)(24) is (143). We write

$$(124) \times (13)(24) = (143).$$

This can be illustrated very clearly by the following diagram where the effect of applying the two permutations in succession can be found by following the arrows.

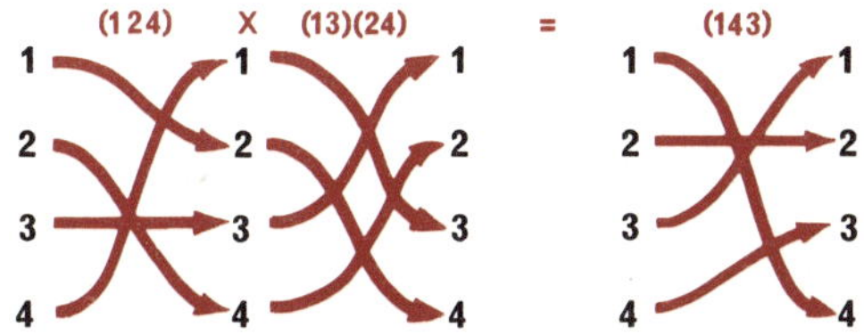

Another example of the same type is the following.

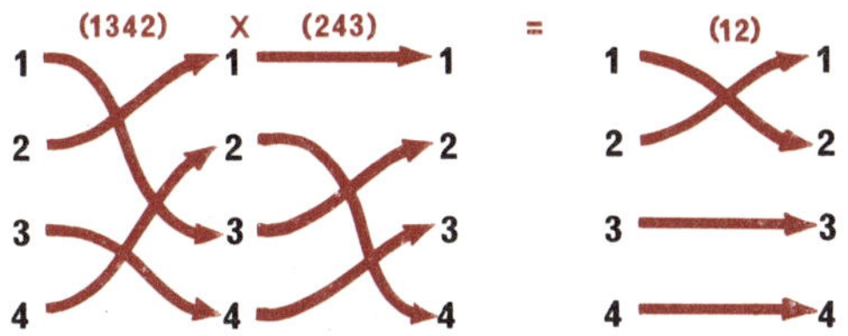

You may find it useful to draw such diagrams to assist you in multiplying permutations. With practice, however, you should be able to dispense with them and write down the product of the two permutations directly.

EXERCISE SET 4

1. Multiply the following permutations:
   (a) (132) × (13) =
   (b) (1423) × (1342) =
   (c) (12)(34) × (123) =
   (d) I × (1253) =
   (e) (12) × (34) =

2. Carry out the following multiplications:
   (a) (12) × (13) =
   (b) (13) × (12) =
   Note that the two answers are not the same.

3. When (1423) is multiplied by one of the following permutations the result is the identity permutation I. Which is it?
   (14)(23), (1234), (12)(43), (1324), I

4. Complete the following multiplication table for S_3.

	I	(123)	(132)	(12)	(13)	(23)
I		(123)	(132)			
(123)				(23)		(13)
(132)		I		(13)		
(12)	(12)				(123)	
(13)				(132)		(123)
(23)	(23)				(132)	

Do permutations commute?

Multiplication of permutations is similar in many ways to the multiplication of numbers. However, there are some important differences. One of these was pointed out in question 2 in the above set of exercises. It is possible to obtain a different answer for the product of two permutations if they are multiplied in the opposite order. For example,

$$(132) \times (12) = (13)$$

but

$$(12) \times (132) = (23)$$

If two mathematical objects give the same answer when multiplied in either order we say that they *commute*. Numbers commute with one another. For example, $4 \times 7 = 28 = 7 \times 4$. In the algebra of numbers, $xy = yx$ always holds. This is called the *commutative law.* However, the commutative law does *not* hold in the algebra of permutations. This does not mean that it is impossible for permutations to commute. For example, (1234) × (13)(24) = (1432) = (13)(24) × (1234). And of course the identity permutation, I, commutes with all permutations. But it is the exception and not the rule for permutations to commute.

Because the commutative law does not hold for permutations we are not at liberty to change the order of the factors in a product unless we have very good reason to. This means that many of the formulae with which we are familiar for numbers do not hold for permutations.

A very good example of this is the formula

$$(xy)^2 = x^2 y^2$$

Written out in full this becomes

$$xyxy = xxyy$$

Now if we are permitted to interchange factors it is clear that these are equal. But without the commutative law there is no reason why they should be equal. In fact if x and y are permutations $(xy)^2$ will frequently be different to $x^2 y^2$. For example, if $x = (132)$ and $y = (12)$, $xy = (13)$ and so $(xy)^2 = \mathrm{I}$. However, $x^2 = (123)$ and $y^2 = \mathrm{I}$, whence $x^2 y^2 = (123)$.

For the same reason, although $(xy)^3 = x^3 y^3$, $(xy)^4 = x^4 y^4$ etc. are always true for numbers, they are frequently false for permutations. The failure of the commutative law makes the algebra of permutations more difficult than the algebra of numbers: an expression such as $a^4 b^3 c b^2 c^3 a^2 c b^3$ cannot be simplified to $a^6 b^8 c^5$ in the algebra of permutations as it could in the algebra of numbers.

The inverse of a permutation

The *inverse* of a permutation is that permutation which reverses the effect of the permutation. If a permutation replaces 1 by 3, its inverse replaces 3 by 1. In terms of diagrams depicting a permutation by means of arrows, a permutation is changed into its inverse by reversing the arrows.

For example, the inverse of the permutation $1 \to 3$, $2 \to 5$, $3 \to 2$, $4 \to 1$, $5 \to 4$ is the permutation $1 \leftarrow 3$, $2 \leftarrow 5$, $3 \leftarrow 2$, $4 \leftarrow 1$, $5 \leftarrow 4$ which is written $3 \to 1$, $5 \to 2$, $2 \to 3$, $1 \to 4$, $4 \to 5$ or $1 \to 4$, $2 \to 3$, $3 \to 1$, $4 \to 5$, $5 \to 2$.

Since the inverse of a permutation reverses its effect the result of following a permutation by its inverse is to leave the objects fixed. Thus if a permutation and its inverse are multiplied together (in either order) the product is the identity permutation, I. This is illustrated by an example in the following diagram.

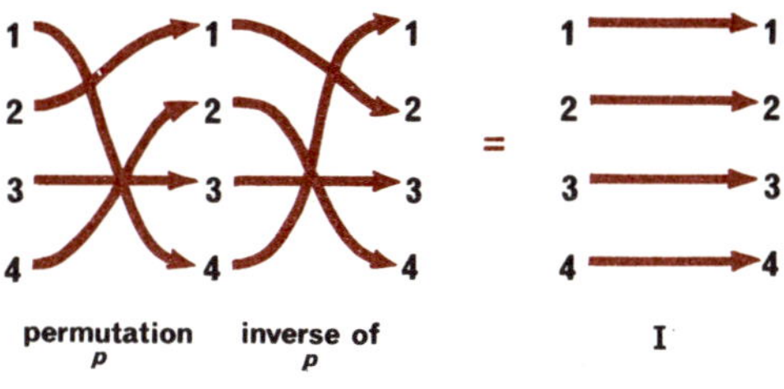

The inverse of the permutation p is denoted by p^{-1}. The observation above can be expressed by the following equation which holds for all permutations p.

$$p\,p^{-1} = \mathrm{I} = p^{-1}p$$

Since the problem of finding the inverse of a permutation involves reversing the direction of the changes that take place, we can find the inverse of a permutation in cycle notation by simply writing it down backwards. The inverse of (134)(25) is (52)(431), which should be rewritten as (143)(25) in accordance with the uniqueness rules.

Examples

$$(1342)^{-1} = (2431) = (1243)$$
$$[(12)(34)]^{-1} = (43)(21) = (12)(34)$$
$$[(1234)(567)]^{-1} = (4321)(765) = (1432)(576)$$
$$(13)^{-1} = (31) = (13)$$

As can be seen from the above examples, some permutations are equal to their own inverses. The identity permutation, too, is its own inverse.

EXERCISE SET 5

1. Find the inverse of the following permutations.
   (a) (154)(263) (d) (12)(34)(56)
   (b) (17246)(35) (e) (1243)
   (c) I (f) (1342)

2. If $x = (253)(46)$ and $y = (126)$, calculate the following.
   (a) xy (d) y^{-1}
   (b) $(xy)^{-1}$ (e) $x^{-1}y^{-1}$
   (c) x^{-1} (f) $y^{-1}x^{-1}$
   Which of the above are equal?

3. State whether the following are true or false.
   (a) $\mathrm{I}^{-1} = \mathrm{I}$
   (b) $(xy)^{-1} = x^{-1}y^{-1}$ for all permutations x and y.
   (c) The inverse of the inverse of a permutation is equal to that permutation itself.
   (d) If $p = p^{-1}$ then $p^2 = \mathrm{I}$.

The algebra of permutations

We have already noted some similarities and some differences between the algebra of numbers and the algebra of permutations. These, and other points of similarity and dissimilarity are listed below.

Similarities

(1) Just as there is an identity number, the number 1, there is the identity permutation, I.

(2) Permutations have inverses just as numbers do. (There is a small difference here, however, because *every* permutation has an inverse while there is one number, the number 0, which does not have an inverse.)

(3) The associative law holds for numbers and permutations alike. This is the law that states that

$$x(yz) = (xy)z$$

for all x, y and z. It is because of this law that we can remove parentheses in a product without ambiguity. For example if $x = (1352)$, $y = (124)(56)$ and $z = (34)$ we can calculate xyz in either of two ways:

$xy = (13654)$, $z = (34)$	$x = (1352)$, $yz = (1234)(56)$
$(xy)z = (14)(365)$	$x(yz) = (14)(365)$

In either case we obtain the same answer, namely that permutation which is equivalent to operating the three permutations x, y and z, one after the other.

A complex product such as $xyxzy$ can be calculated in many different ways, including $[(xy)(xz)]y$, $[x(yx)](zy)$ and $[x(y(xz))]y$. The associative law ensures that the result will always be the same.

(4) We can carry out cancellation in the algebra of permutations in much the same way as we can for numbers. There are some differences however. If $xy = xz$ we can cancel the x's as follows:

$$\begin{aligned} xy &= xz \\ x^{-1}xy &= x^{-1}xz \\ \mathrm{I}y &= \mathrm{I}z \\ y &= z \end{aligned}$$

We must be careful, however, that we cancel factors on the same side of the expression: both must be on the left or both on the right. There is not much we can do with $xy = zx$. We cannot cancel the x's. The best we could do is to multiply both sides of the equation by x^{-1} on the left and obtain $y = x^{-1}zx$.

(5) It is true both for numbers and for permutations that $(x^{-1})^{-1} = x$ (the inverse of the inverse of x is x itself).

(6) It is true both for numbers and permutations that $(xy)^{-1} = y^{-1}x^{-1}$ since $xyy^{-1}x^{-1} = x\mathrm{I}x^{-1} = xx^{-1} = \mathrm{I}$. The inverse of a product is the product of the inverses *in reverse order.* This extends to products of any number of factors. For example, $(abcd)^{-1} = d^{-1}c^{-1}b^{-1}a^{-1}$.

Differences

(1) It is possible to add numbers as well as to multiply them. Permutations can only be multiplied. Expressions such as $x + y$ and $-x$ which relate to addition have no meaning for permutations. Moreover there is no permutation corresponding to the number '0'.

(2) In the algebra of numbers the commutative law, $xy = yx$, holds whereas in the algebra of permutations it does not.

(3) $(xy)^2 = x^2y^2$, $(xy)^3 = x^3y^3$ etc. are valid for all numbers x, y and z. However, they do not, as a rule, hold for permutations.

(4) $(xy)^{-1} = x^{-1}y^{-1}$ holds for numbers but not, as a rule, for permutations. However see (6) in the list of similarities above.

(5) $x^{-1}yx = y$ holds for numbers but not usually for permutations. A permutation cannot be cancelled with its inverse unless they are adjacent factors.

(6) There are only two solutions to the equation $x^2 = 1$ in the algebra of numbers, $x = \pm 1$. However there is any number of solutions to the equation $x^2 = \mathrm{I}$ in the algebra of permutations. For example (12), (13), (14), (23), (12)(34), (12)(34)(56) etc. all produce the identity permutation when squared.

(7) If $x^2 = y^2$ in the algebra of numbers we may conclude that $x = \pm y$. However there is no simple conclusion comparable to this if x and y are permutations. For example, the squares of (123456), (143652), (153)(264) and (123456)(78) and many other permutations are all equal to (135)(246).

Powers

Just as for numbers we define powers of a permutation. If p is a permutation, p^1 is just p, p^2 is pp, p^3 is ppp and so on. Adding 1 to the index of p has the effect of multiplying by p. Subtracting 1 from the index has the effect of multiplying by the inverse of p. For example, $p^2 = p^3 p^{-1}$, $p^1 = p^2 p^{-1}$. Continuing, we must have $p^0 = p^1 p^{-1} = \mathrm{I}$, $p^{-1} = p^0 p^{-1} = \mathrm{I}p^{-1} = p^{-1}$ (the reason for writing the inverse as the minus first power is so that it will fit into this pattern). Continuing further, $p^{-2} = p^{-1}p^{-1} = (p^{-1})^2$, $p^3 = p^{-2}p^{-1} = (p^{-1})^2 p^{-1} = (p^{-1})^3$, and so on.

With one exception, the usual index laws for numbers hold also for permutations. They are

(1) $a^\circ = \mathrm{I}$	(3) $(a^m)^n = a^{mn}$
(2) $a^m . a^n = a^{m+n}$	(4) $a^{-n} = (a^{-1})^n$

These four laws hold for all permutations, a, and all whole numbers, n (positive, negative or zero). The index law that does *not* hold for permutations is $(ab)^n = a^n b^n$.

Raising permutations to a power when the permutations are expressed in cycle notation is very easy. We simply need to remember that raising a permutation to a power is equivalent to repeating the permutation a certain number of times.

Examples

(1) $(14325)^2 = (13542)$

We move forward through the cycle taking two steps at a time: $1 \to 4 \to 3$, $3 \to 2 \to 5$, $5 \to 1 \to 4$, $4 \to 3 \to 2$, $2 \to 5 \to 1$.

(2) $(14325)^3 = (12453)$

(3) $(14325)^4 = (15234)$

(4) $(14325)^5 = \mathrm{I}$

(5) $(143256)^2 = (135)(264)$

(6) $(143256)^3 = (12)(36)(45)$

(7) $[(1432)(567)]^2 = (1432)^2(567)^2 = (13)(24)(576)$

In the last example above, we claimed that $(ab)^2 = a^2 b^2$ where $a = (1432)$ and $b = (567)$. Since the equation $(ab)^2 = a^2 b^2$ is not generally valid for permutations we must have good reason for using it. The reason is that a and b commute in this case. $ab = ba = (1432)(567)$. This will always occur when a permutation involves two or more cycles: the cycles commute with one another since they involve different objects.

Examples

(8) $[(123)(45)(6789)]^2 = (123)^2(45)^2(6789)^2 = (132)\mathrm{I}(68)(79)$
$= (132)(68)(79)$

(9) $[(123)(45)(6789)]^6 = [(132)(68)(79)]^3 = (132)^3(68)^3(79)^3$
$= \mathrm{I}(68)(79) = (68)(79)$

(10) $[(123)(45)(6789)]^{12} = [(68)(79)]^2 = (68)^2(79)^2 = \mathrm{II} = \mathrm{I}$

Order of a permutation

The above examples illustrate what happens to a permutation when it is raised to higher and higher powers. Eventually the identity permutation is reached and from then on the sequence of powers is repeated. For example, if $x = (123)(45)$, $x^2 = (132)$, $x^3 = (45)$, $x^4 = (123)$, $x^5 = (132)(45)$, $x^6 = \mathrm{I}$, $x^7 = (123)(45) = x$, $x^8 = x^2, \ldots$ If x^n is the first power of x (for positive n) that is equal to I, we say that n is the *order* of the permutation x. The order of (123)(45) is thus 6.

The order of the single cycle $(x_1, x_2, \ldots, x_n)$ is simply n since it requires n steps for each object to revert to its original position. The order of a permutation which is made up of two or more cycles is the least common multiple of the orders of the cycles. For example, if a permutation, p, is made up of two cycles, one containing 4 objects and the other containing 6 objects, in order to reduce the first cycle to the identity we must raise p to a power which is a multiple of 4 and to reduce the second cycle to the identity we must raise p to a multiple of 6. The first time that the two cycles will be reduced to the identitv simultaneously (and the whole permutation can only be reduced to the identity if each cycle is reduced, since the cycles permute different objects) will be in the 12th power.

This illustrates one of the many advantages of the cycle notation. The order of a permutation can be very simply found.

> *The order of a permutation is the least common multiple of the lengths of its cycles.*

EXERCISE SET 6

1. State which of the following are meaningless, which are always true, and which can be false, in the algebra of permutations.
   (a) $a^2 - b^2 = (a-b)(a+b)$
   (b) $x^{-1}y^{-1}xy = \mathrm{I}$

(c) $x = x^{-1}$
(d) $(x^{-1}y^{-1})^{-1} = yx$
(e) every permutation commutes with its inverse
(f) if $x^3 = I$ then $x = I$

2. If $x = (1342)(567)$ write down the positive powers of x until they begin to repeat. What is the order of x?

3. Write down the orders of the following permutations.
(a) (12345)
(b) (12345)(67)
(c) (123456)(78)
(d) (12)(34)(56)(78)
(e) I

4. Write down all of the permutations in S_4 which have order 2.

5. In each of the following cases find the order of xy.
(a) $x = (123)$, $y = (12)$
(b) $x = (123)$, $y = (45)$
(c) $x = (123)$, $y = (14)$
(d) $x = (123)$, $y = (12)(34)$
(e) $x = (123)(456)$, $y = (34)(56)$
(Note that in all cases, x has order 3 and y has order 2, while the order of xy is different in all five cases. This shows that there is no connection between the order of a product and the order of the factors.)

6. If x is a permutation of order 8 show that
(a) x^2 has order 4
(b) x^6 has order 4
(c) x^5 has order 8.

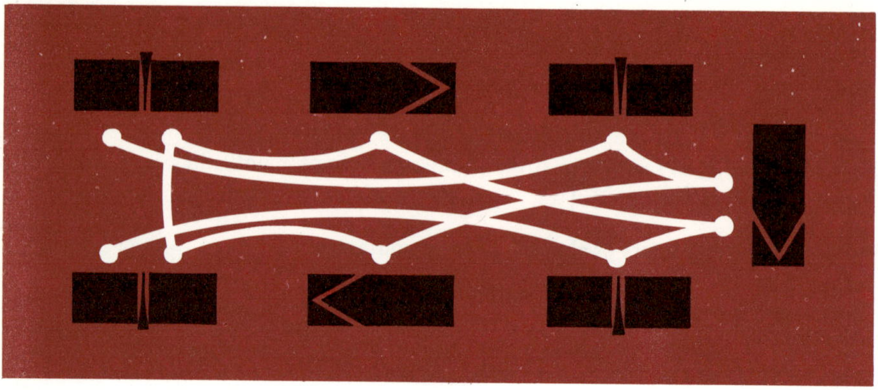

Permutation Problems

Problems in a toothpick factory

Once there was a toothpick factory which employed seven men. There were seven machines – four cutting machines, for cutting pieces of wood to size, and three sharpening machines for giving the toothpicks their fine points. These machines were arranged around the walls of the small factory as indicated in the following floor-plan.

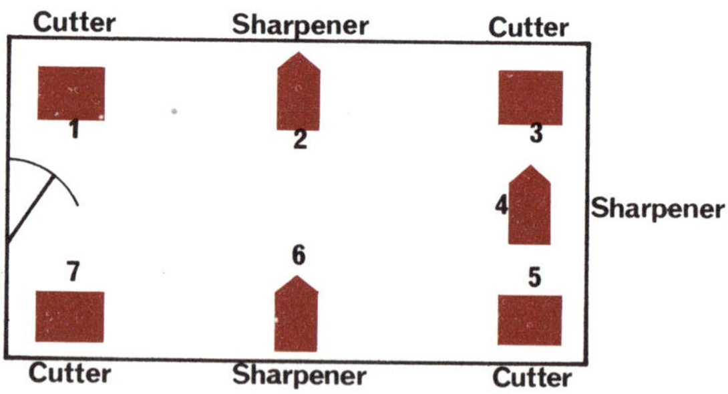

The owner of the factory found that the men worked best if they changed machines frequently and so he drew up the following arrangements for the workers to move from one machine to another.

The men worked a 5-day week, from Monday to Friday. Every morning they moved to the machine next on the right to the one they had operated on the previous working day. Every afternoon after lunch they moved again to the right, but this time to the next machine of the same type. If they had been working on a cutter in the morning they moved to the next cutter. These arrangements meant that the men changed machines every half day, but they worked on the same type of machine for a whole day.

For example, if a worker had worked on machine 1 (cutter) on Monday afternoon, he would move to machine 2 (sharpener) on Tuesday morning (i.e. in more precise permutation terms, for him machine 1 would be 'replaced by' machine 2, or $1 \rightarrow 2$). After lunch on Tuesday he would move to machine 4 (the next sharpener to the right). On Wednesday morning he would go to machine 5 and on Wednesday afternoon to machine 7.

On the wall of the factory was hung a chart which showed who would be working on which machine during each half-day period. This roster had to be changed from week to week because a man who, before the Monday morning change (it is convenient to take this as the starting point), was on say machine 2 would find himself on a different machine at the same time the following week. However, they soon discovered that after some weeks the rosters began to repeat themselves and a chart that had been used some weeks ago was again applicable. The question we now ask is, after how many weeks will the rosters begin to repeat? Of course this can be solved by actually drawing up the rosters, but the question can be most easily solved using permutations.

Let m denote the permutation which corresponds to the change that takes place first thing each morning and let a denote the permutation corresponding to the 'after-lunch' change. Then $m = (1234567)$ and $a = (1357)(246)$. The permutation ma will describe the change that takes place from one day to the next and $(ma)^5$ describes the change that takes place from the end of one week to the end of the next. Putting $w = (ma)^5$, w^2 describes the change that takes place after two weeks, w^3 describes what happens after three weeks, and so on. As we raise w to higher and higher powers we know that eventually we will obtain a power of w, say w^n, which is equal to the identity permutation. This means that after n complete weeks the rosters will begin to repeat. This n is simply the order of w.

$$m = (1234567), \qquad a = (1357)(246)$$
$$ma = (14736)(25)$$
$$w = (ma)^5 = (14736)^5(25)^5 = (25)$$

The order of w is 2 and so the rosters will repeat themselves every two weeks. That is, the owner of the factory need only draw up two different weekly roster charts which may be used in alternate weeks.

Over the years, circumstances changed and the demand for toothpicks increased. The owner of the factory decided to operate it on a 5½-day week by paying the men overtime to work on Saturday morning. This meant that the rosters had to be changed. This time, $w = (ma)^5 m = (14736)^5 (25)^5 \times (1234567) = (25) \times (1234567) = (1267)(345)$. The order of w is now 12 (the least common multiple of the lengths of the cycles) and so under these new arrangements, the rosters will not begin to repeat until 12 weeks have elapsed.

After working the 5½-day week for some time, the men asked if they could work on Saturday afternoons instead of Wednesday mornings so that they could get to the shops. This request was granted, but it meant another change to the rosters. This time, $w = (ma)^2 a(ma)^3 = (17643) \times (1357)(246) \times (13467)(25) = (1342675)$. The order of w is now 7, so under these new arrangements the rosters will repeat after only 7 weeks.

EXERCISE SET 7

1. Suppose that the toothpick factory works all day Monday to Thursday, Friday mornings, and all day Saturday. After how long will the weekly rosters begin to repeat?

2. Improvements to the sharpening machines have meant that one worker has been put off and machine 6 has been closed down. Bert Thomas is working on machine 3 one Monday morning. When will he next be working on that machine on a Monday morning assuming
   (a) the factory works a 5½-day week (Monday to Friday plus Saturday morning)
   (b) the factory works a 5-day week (Monday to Friday)?

Card shuffling

The purpose of shuffling a pack of cards before dealing is to rearrange the cards in a random manner so that no player knows which cards are being dealt to his opponents. However, if the shuffling is done in a controlled manner it is possible, using permutations, to predict the final arrangement.

To illustrate this, we consider a pack of eight cards. It is suggested that the reader removes the first eight cards in any suit (ace to eight) and follows the shuffling procedures as they occur. In what follows we refer to the ace by the numeral 1.

Take the cards in order, face up, with 1 on top. Now deal them alternately into two piles face down beginning with the left-hand pile. Turn over the piles and place the left-hand pile on top of the right-hand one. The order of the cards should now be 1,3,5,7,2,4,6,8. We call this shuffling operation '*d*'. As a result of this operation the card in position 1 is unchanged, the card which was previously in position 2 goes to position 5 and so on. The change of positions can be described by 1→1, 2→5, 3→2, 4→6, 5→3, 6→7, 7→4, 8→8 and so regarding this as a permutation of the positions, $d = (253)(467)$.

It may be objected that *d* changes the arrangement of the cards from 1,2,3,4,5,6,7,8 to 1,3,5,7,2,4,6,8 and that therefore it corresponds to the permutation 1→1, 2→3, 3→5, 4→7, 5→2, 6→4, 7→6, 8→8 which is (235)(476). However the operation does not take account of the numbers on the cards but only of the positions the cards occupy. For example position 5 is replaced by position 3 (5→3), so the card in that position (card 5) 'moves' from 5 to 3.

Put the cards back into order and carry out operation '*d*' on them three times. You should find that the cards have reverted to their original order. That is, performing '*d*' three times in succession is equivalent, as far as the final effect is concerned, to doing nothing at all. What you have just done is to verify that $d^3 = I$.

There are three further operations we wish to consider, and to bring out their differences and similarities we describe all four operations in the following table.

$d = (253)(467)$	$l = (147)(285)$	$r = (18)(243756)$	$c = (15)(26)(37)(48)$
take pack with cards face up	take pack with cards face up	take pack with cards face up	take pack with cards face up
deal cards alternately into two piles face *down*, starting with left-hand pile	deal cards alternately into two piles face *up*, starting with left-hand pile	deal cards alternately into two piles face *up*, starting with left-hand pile	take top four cards and put on bottom of pack
turn each pile over			
put *left*-hand pile on top of other pile	put *left*-hand pile on top of other pile	put *right*-hand pile on top of other pile	

Perform each of the operations l, c and r on your pack of cards and check that they correspond to the permutations given in the top line of the table. Do not forget that the permutation describes the change in the positions and not the change in the card numbers for given positions. (The card which was in position x is now in position y, so $x \to y$.)

We may now predict the effect of performing these operations successively by calculating the product of the corresponding permutations. If the operation d is performed, followed by the operation l, the result is described by the permutation $dl = (253)(467) \times (147)(285) = (146)(385)$. If operation r is performed six times in succession the cards will revert to their original positions, since $r^6 = \mathrm{I}$.

Take your eight cards and put them face up in the order 1,2,3,4,5,6, 7,8 with 1 on top. Carry out the operations d, l and r successively, in that order, now repeat operations l and r and finally perform operation l. You should find that your cards have returned to the order 1,2,3,4,5, 6,7,8. This could have been predicted by calculating $dlrlrl = d(lr)^2 l$

$$
\begin{aligned}
&= (253)(467) \times [(147)(285) \times (18)(243756)]^2 \times (147)(285) \\
&= (253)(467) \times [(137862)(45)]^2 \times (147)(285) \\
&= (253)(467) \times (176)(382) \times (147)(285) \\
&= (253)(467) \times (235)(476) \\
&= \mathrm{I}
\end{aligned}
$$

If you consider the difference between the operations l and r you will see that this difference can be compensated by applying the operation c. Thus $r = lc$ (and similarly $l = rc$). Another connection between the operations is that $r^4 = (253)(467) = d$.

Once some relationships between certain permutations have been discovered, it is often easy to discover other relationships by working in the *algebra* of permutations rather than in the *arithmetic* of permutations. The following is an example.

Now $d(lr)^2 l = \mathrm{I}$ and $d = r^4$. Combining these, $r^4 (lr)^2 l = \mathrm{I}$. Thus $rrrrlrlrl = \mathrm{I}$ and so $r^3 (rl)^3 = \mathrm{I}$. Multiplying each side of this equation on the left by r^3, and using the fact that $r^6 = \mathrm{I}$, we obtain the equation $(rl)^3 = r^3$. Substituting $r = lc$, we have $(lcl)^3 = (lc)^3$ and so $lcllcllcl = lclclc$. Cancelling lcl at the left of each side of this equation (i.e. multiplying each side on the left by $l^{-1} c^{-1} l^{-1}$) we have $lcl^2 cl = clc$. Multiplying by c on the left and on the right of each side, $clcl^2 clc = c^2 lc^2 = l$, since $c^2 = \mathrm{I}$. Hence $clcl^2\ clcl^2 = l^3 = \mathrm{I}$ and so $(clcl^2)^2 = \mathrm{I}$.

EXERCISE SET 8

1. A new operation, f, on the eight cards is defined as follows:
   'Take the pack with the cards face up. Deal the cards one at a time into four piles, face up, starting with the left-hand pile. Stack these piles on top of one another with the left-hand pile on top, the second pile underneath it and so on.' Carry out this operation on your pack of cards and write down the permutation that corresponds to it.

2. Find a relationship between f and l and use it to express d in terms of f and c.

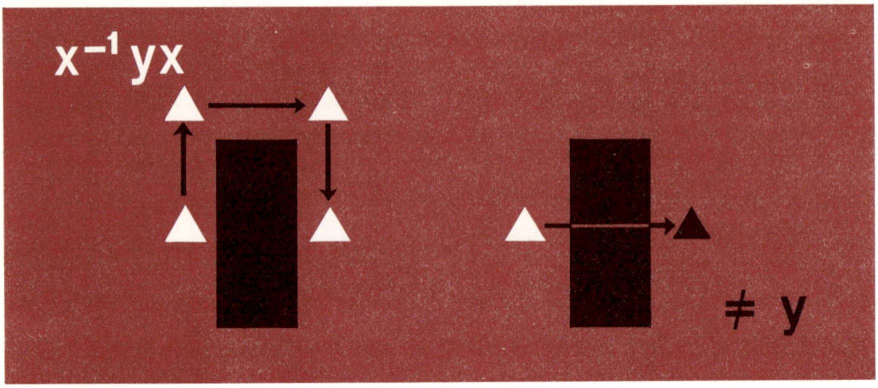

Conjugation

A new operation

If x and y are numbers, the expression $x^{-1}yx$ can be simplified to $x^{-1}xy = y$ since x and y commute (i.e. $yx = xy$). However if x and y are non-commuting permutations, $x^{-1}yx$ will be different from y. For example if $y = (12)(345)$ and $x = (12345)$, $x^{-1}yx = (15432)(12)(345)(12345) = (145)(23)$. This is a new operation called *conjugation.* $x^{-1}yx$ is called a *conjugate* of y.

We denote the expression $x^{-1}yx$ by y^x. We write conjugation in the same way as we do powers for numbers because some of its properties are similar. However it must be remembered that if x and y are permutations, y^x is *not* the operation of raising y to the xth power or of multiplying y by itself x times: these are meaningless where x is a permutation.

The properties of the operation of raising a number to a power are summarized in the following table of index laws.

Index laws for numbers

(1) $(ab)^c = a^c b^c$
(2) $(a^b)^c = a^{bc}$
(3) $a^1 = a$
(4) $(a^{-1})^b = (a^b)^{-1}$

} True for conjugation of permutations

(5) $a^b a^c = a^{b+c}$ (6) $a^{-b} = (a^b)^{-1}$ (7) $a^0 = 1$	Meaningless for conjugation of permutations

Laws (5), (6) and (7) have no meaning since there are no operations of addition and subtraction for permutations, and there is no 'zero' permutation.

Laws (1) to (4) hold for conjugation of permutations:

(1) $(ab)^c = a^c b^c$

$a^c b^c = c^{-1}ac \cdot c^{-1}bc = c^{-1}a(cc^{-1})bc = c^{-1}abc = (ab)^c$

(2) $(a^b)^c = a^{bc}$

$(a^b)^c = c^{-1}(a^b)c = c^{-1}b^{-1}abc = (bc)^{-1}abc = a^{bc}$

(3) $a^{\mathrm{I}} = a$

$a^{\mathrm{I}} = \mathrm{I}^{-1}a\mathrm{I} = \mathrm{I}a\mathrm{I} = a$

(4) $(a^{-1})^b = (a^b)^{-1}$

$(a^b)^{-1} = (b^{-1}ab)^{-1} = b^{-1}a^{-1}b = (a^{-1})^b$

The cycle structure of a conjugate

A conjugate of x need not be equal to x, but it must have something in common with x. What this 'something' is will become apparent from examining the following examples.

$x = (12)$	$y = (14)(235)$	$x^y = (34)$
$x = (123)$	$y = (1652)$	$x^y = (136)$
$x = (14)(25)$	$y = (34)$	$x^y = (13)(25)$
$x = (12476)$	$y = (13)(27)$	$x^y = (26374)$

A conjugate of a permutation always has the same cycle structure as that permutation. For example, every conjugate of a 3-cycle is a 3-cycle: every conjugate of a permutation consisting of three 2-cycles, a 3-cycle and two 4-cycles is also made up from three 2-cycles, a 3-cycle and two 4-cycles. Moreover, one can find the result of conjugating a permutation x by a permutation y by applying y to the symbols in the cycle notation for x. For example, if $x = (12)$ and $y = (14)(235)$, $x^y = (43)$ since under the action of y, $1 \to 4$ and $2 \to 3$. If $x = (12476)$ and $y = (13)(27)$, $x^y = (37426)$ since under the action of y, $1 \to 3$, $2 \to 7$, $4 \to 4$, $7 \to 2$ and $6 \to 6$. As in these two cases, we usually have to rewrite the permutation according to the uniqueness principles of chapter 1. In this last example, we rewrite (37426) as (26374).

Not only does a conjugate of a permutation have the same cycle structure as the permutation, but any two permutations which have the same cycle structure are conjugates of one another. For example, $x = (14)(265)$ and $z = (132)(46) = (46)(132)$ have the same cycle structure. We seek a suitable permutation y so that $x^y = z$. We require that under the action of y, $1\to4$, $4\to6$, $2\to1$, $6\to3$, $5\to2$. This does not stipulate the action of y on 3. We are free to let 3 be replaced by any symbol apart from 4,6,1,3 and 2. Of course, if we are considering permutations on the symbols 1,2,3,4,5,6 only this does not leave much choice: we must have $3\to5$. However if the symbols 7,8,9, . . . are available we may have $3\to7$ or $3\to8$ or $3\to9$, . . . Suppose we let $3\to5$ under the action of y. Then y is the permutation $1\to4$, $4\to6$, $2\to1$, $6\to3$, $5\to2$, $3\to5$ which is written (146352). Check for yourself that $(146352)^{-1}\ (14)(265)(146352) = (132)(46)$.

The conjugating permutation, y, in the above example is far from being unique. There are many correct possibilities. For example we could write x as (41)(265) and z as (46)(213). We would now seek a permutation y which has the effect $4\to4$, $1\to6$, $2\to2$, $6\to1$, $5\to3$. Suppose we again let $3\to5$. This time we take $y = (16)(35)$. Note that while conjugates must have the same cycle structure, conjugating elements which produce the same conjugate need not. They might be as different as (146352) and (16)(35).

Example 1. If $x = (1342)(57)$ and $y = (265)$, find x^y.

Solution. $x^y = (1346)(27)$

Example 2. If $x = (1342)$ and $z = (1234)$, find y so that $x^y = z$.

Solution. We seek y so that under its action, $1\to1$, $3\to2$, $4\to3$, $2\to4$. Take $y = (243)$.

Example 3. If $x = (146)$ and $z = (237)$, find y so that $x^y = z$.

Solution. We seek y so that under its action, $1\to2$, $4\to3$, $6\to7$. It remains to stipulate the action of y on 2,3,5 and 7. We may send these to any symbols apart from 2,3, and 7 which have already been used. We could stipulate that $2\to4$, $3\to1$, $5\to6$ and $7\to5$. This results in a permutation $y = (1243)(567)$. Or we could insist that $2\to1$, $3\to4$, $5\to5$ and $7\to6$. This time y would be (12)(34)(67). There are in fact many different but correct answers.

EXERCISE SET 9

1. Find x^y in the following cases.

   (a) $x = (12)$ $\quad y = (2345)$

   (b) $x = (13)(256)$ $\quad y = (23)(45)$

   (c) $x = (14)(25)$ $\quad y = (1325)$

   (d) $x = (1325)$ $\quad y = (14)(25)$

   (e) $x = (156)$ $\quad y = (156)$

2. Find a permutation, y, so that $x^y = z$ where this is possible in the following cases:

   (a) $x = (132)(45)$ $\quad z = (15)(234)$

   (b) $x = (14)(23)$ $\quad z = (16)(24)(35)$

   (c) $x = (12)(34)$ $\quad z = (13)(24)$

   (d) $x = (14)$ $\quad z = \mathrm{I}$

   (e) $x = (12)$ $\quad z = (13)$

3. Simplify the following:

   (a) (12)(13)

   (b) (12)(13)(14)

   (c) (12)(13)(14)(15)(16)(17)

Breaking down cycles

We are familiar with the fact that every permutation can be broken down into a product of disjoint cycles. ('Disjoint' means that no symbol appears in two different cycles.) The whole system of writing permutations in cycle notation is built on this fact. We can, however, carry the process of breaking down a permutation even further by breaking each cycle into cycles of length two (2-cycles). In question 3 in the above set of exercises we saw how this can be done. For example, (1234567) = (12)(13)(14)(15)(16)(17). This time the cycles will no longer be disjoint. In this example the symbol '1' is repeated.

Since any permutation can be factorized into cycles and any cycle can be factorized into 2-cycles, then any permutation can be written as a product of 2-cycles. For example, factorizing (1346)(28)(579) into 2-cycles we obtain (13)(14)(16)(28)(57)(59).

The way a permutation is broken down into 2-cycles is far from being unique. For example, simplifying (14)(24)(13)(25)(23) we obtain (153)(24). Breaking this up into 2-cycles again we obtain (15)(13)(24). Thus (153)(24) can be written as a product of 2-cycles in two quite different ways. Even the number of factors is different. The only significant similarity between the factorizations (14)(24)(13)(25)(23) and (15)(13)(24) is that in each case the number of factors is odd. This feature will be explored further in a later chapter.

Generators of S_n

Imagine an electronic device consisting of two buttons and a screen on which the symbols 1,2,3,4,5,6,7,8 are depicted in some order and suppose that each button when pressed produces a certain permutation of the eight symbols. It would be interesting to know whether every permutation could be produced using only the two buttons.

For example, suppose that button A produces the permutation (13276) and button B produces the permutation (18)(256). If the arrangement of symbols on the screen is 1,2,3,4,5,6,7,8, can it be changed to 8,6,7,4,1,2,5,3 using only these two buttons? The permutation required to carry out this change is (15738)(26). (What was in position 1 is now in position 5 etc.) The question is whether (15738)(26) can be built up from the two permutations A = (13276) and B = (18)(256). (No confusion will arise by using the same symbols for the buttons as for the permutations produced by them.) The answer is that it can, for it can be shown that

$$(15738)(26) = A^2BAB^2A^4BAB.$$

This means that the desired effect can be obtained by pressing the buttons according to this pattern – press button A twice, then button B, button A, button B twice etc.

If a permutation, h, can be built up from two permutations A and B in this way, we say that A and B *generate* h. What was shown above is that (13276) and (18)(256) generate (15738)(26). However not every permutation can be generated by (13276) and (18)(256). For example (14) cannot be generated by them. This is because (13276) and (18)(256) both fix the symbol 4 and so every permutation that can be built up from them must fix 4. No such permutation can possibly interchange 1 and 4. In terms of the permutation machine, neither button has any effect on the symbol in the fourth place.

Can we construct a permutation machine such that *every* arrangement can be achieved by using only the two buttons? To do this we must find two permutations A and B such that every permutation on the eight symbols can be built up from them. In other words we must find two permutations A and B which generate every permutation in S_8. There are many possible choices for A and B.

(12) and (234 . . . n) generate every permutation in S_n

We present below a general method for generating permutations in S_n by $A = (12)$ and $B = (234 \ldots n)$ together with the application of this method to a particular example.

General method:

To generate the permutation, p, by $A = (12)$ and $B = (234 \ldots n)$ carry out the following steps.

(1) Write p as a product of 2-cycles.
(2) Replace each of those 2-cycles (rs), which do not contain the symbol 1 by $(1r)(1s)(1r)$. Note: $(rs) = (1r)(1s)(1r)$. p will now be written as a product of 2-cycles of the form $(1t)$.
(3) Replace each 2-cycle $(1t)$ by $B^{-(t-2)} AB^{t-2}$. Note: B^{t-2} produces the effect $2 \to 3 \to \ldots \to t$ and $1 \to 1 \to \ldots \to 1$. Thus $(12)^{B^{t-2}} = (1t)$, i.e. $B^{-(t-2)} AB^{t-2} = (1t)$.
(4) Use the fact that $B^{n-1} = I$ to replace negative powers of B by positive ones. For example $B^{-1} = B^{n-2}$.

Example. $n = 8$, $A = (12)$, $B = (2345678)$, $p = (1437)(258)$

(1) $p = (14)(13)(17)(25)(28)$
(2) $= (14)\cdot(13)\cdot(17)\cdot(12)(15)(12)\cdot(12)(18)(12)$
$= (14)(13)(17)(12)(15)(18)(12)$ since $(12)(12) = I$
(3) $= B^{-2} AB^2 \cdot B^{-1} AB \cdot B^{-5} AB^5 \cdot A \cdot B^{-3} AB^3 \cdot B^{-6} AB^6 A$
$= B^{-2} ABAB^{-4} AB^5 AB^{-3} AB^{-3} AB^6 A$
(4) $= B^5 ABAB^3 AB^5 AB^4 AB^4 AB^6 A$ since $B^7 = I$

(12) and (1234 . . . n) generate every permutation in S_n

Put $A = (12)$, and $C = (1234 \ldots n)$.

Then $B = CA = (1234 \ldots n)(12) = (234 \ldots n)$.

To generate any permutation, p, by A and C simply follow the above method for generating p by A and B and replace B by CA wherever it appears. For example, if $n = 8$ and $p = (1437)(258)$ we saw above that

$$\begin{aligned} p &= B^5 ABAB^3 AB^5 AB^4 AB^4 AB^6 A \\ &= (CA)^5 A(CA)A(CA)^3 A(CA)^5 A(CA)^4 A(CA)^4 A(CA)^6 A \\ &= CACACACACA^2 CA^2 CACACA^2 CACACACACA^2 CACACACA^2 \times \\ &\quad CACACACACACA^2 \end{aligned}$$

Although the above methods give systematic ways for generating any permutation they do not necessarily give the simplest generating formula. For example, if $A = (12)$, $B = (234567)$ and $C = (1234567)$ then $CAC = (1246)(357)$. So (1246)(357) can be generated very simply by A and C. However, if we carry out the above method we obtain

$$
\begin{aligned}
(1246)(357) &= (12)(14)(16)(35)(37) \\
&= (12)(14)(16)(13)(15)(13)(13)(17)(13) \\
&= (12)(14)(16)(13)(15)(17)(13) \\
&= AB^{-2}AB^{2}\,B^{-4}AB^{4}B^{-1}ABB^{-3}AB^{3}\,B^{-5}AB^{5}B^{-1}AB \\
&= AB^{-2}AB^{-2}AB^{3}AB^{-2}AB^{-2}AB^{4}AB \\
&= AB^{4}AB^{4}AB^{3}AB^{4}AB^{4}AB^{4}AB \\
&= A(CA)^{4}A(CA)^{4}A(CA)^{3}A(CA)^{4}A(CA)^{4}A(CA)^{4}A(CA) \\
&= ACACACACA^{2}CACACACA^{2}CACACA^{2}CACACACA^{2} \times \\
&\quad\; CACACACA^{2}CACACACA^{2}CA
\end{aligned}
$$

which is considerably more complicated than CAC!

EXERCISE SET 10

1. Write (1526)(374) as a product of 2-cycles.

2. Generate (123)(456) by $A = (12)$ and $B = (23456)$.

3. Generate (259)(37) by $A = (12)$ and $C = (123456789)$.

4. Generate (15)(263) by $F = (12)(345)$ and $B = (23456)$.
   [Hint: Express $A = (12)$ as a power of F.]

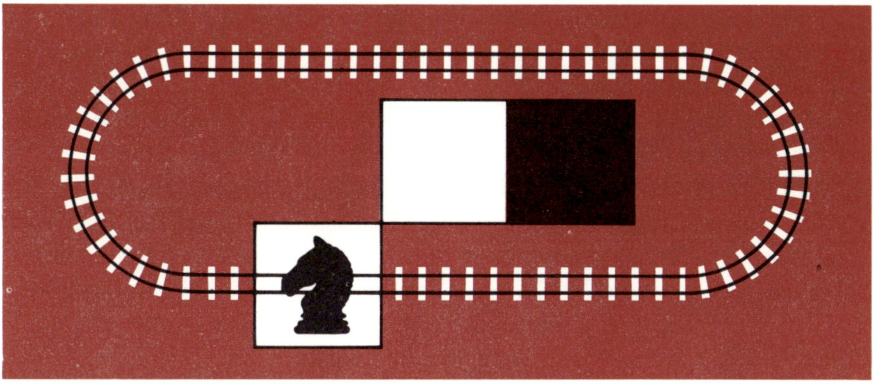

Shunting and Jumping Puzzles

In a certain shunting yard there is a loop of track. On one side of the loop there is a turntable. This turntable can only operate with four carriages. It is not big enough to operate with more than four carriages and for some mysterious reason it will not operate with fewer than four carriages.

On this loop of track there is a train consisting of an engine followed by six carriages and a guard's van on the end. The problem is to interchange the engine and the guard's van while keeping the six carriages in the same order. This must be done using only the loop of track and the turntable. Any number of carriages can be taken around the loop at

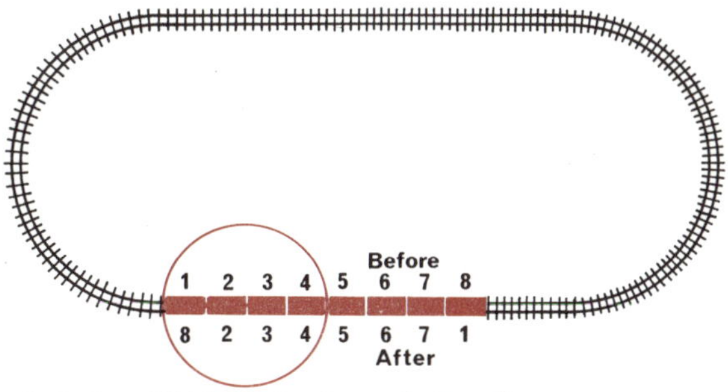

1 = Engine; 2,3,4,5,6,7 = Carriages; 8 = Guard's van

any time and the turntable can be operated at any time so long as it has exactly four carriages on it (the engine and the guard's van are each equivalent to one carriage).

This may not sound like a problem which is likely to arise in a real-life shunting yard but it makes a very interesting puzzle. You can simulate the problem by taking eight coins to represent the engine, guard's van and carriages, pasting little paper labels on them to identify them as such. The 'turntable' can be operated by putting four fingers on four of the 'carriages' and rotating. If you try to solve this puzzle using just inspired trial and error you will perhaps be more impressed by the relative ease with which it can be solved using permutations.

Expressing the problem using permutations

For convenience we use the symbols 2,3,4,5,6,7 to represent the six carriages, the symbol 1 to represent the engine and 8 to represent the guard's van. The initial arrangement of the train is 12345678. The arrangement that we are trying to achieve is 82345671. The permutation which produces this change is (18).

There are two basic operations at our disposal. We can take one carriage from the left-hand side of the train around the loop to the right-hand side (throughout this discussion the term 'carriage' will include the engine and guard's van). This operation corresponds to the permutation l = (18765432) since the carriage which was previously in position 1 ends up in position 8, the carriage which was in position 8 goes to position 7, and so on. Everything that can be done using just the loop can be built up from the permutation l. For example, taking a number of carriages around the loop at the same time is equivalent to taking them one at a time – that is, equivalent to carrying out the operation l the required number of times. Taking one or more carriages from the *right*-hand side across to the left is equivalent to performing the inverse of the permutation one or more times.

Rotating the turntable is equivalent to applying the permutation t = (14)(23). Do not forget that these symbols here refer to the positions. For convenience we have the turntable operating on positions 1,2,3,4 but of course any of the eight carriages can occupy these positions. Moving the train forward so that the carriages in positions 2,3,4,5 can occupy the turntable, and then rotating the turntable, is equivalent (as far as the permutations go) to carrying out permutation l and following by t – that is, performing the permutation lt.

Removing the irrelevant details of the puzzle, we can express it purely in terms of permutations:

Problem: Generate (18) by means of $l = (18765432)$ and $t = (14)(23)$.

Solution of the problem in terms of permutations

$$l = (18765432),\ t = (14)(23)$$

We are trying to generate (18) by means of l and t. Of course, we could try various combinations at random until we hit upon one that works. This is obviously not a very efficient procedure. The method we shall use, although it relies to a certain extent on trial and error, is much more direct.

We want to generate a particular 2-cycle. However if we can generate *any* 2-cycle, there is a chance that we can obtain the one we want by conjugation. For example, if we could generate (48) using l and t, then by conjugating by t we would obtain (18) since

$$(48)^t = (48)^{(14)(23)} = (18).$$

This observation has increased the number of acceptable possibilities. However it might still be difficult to obtain a 2-cycle. But if we were able to find a permutation made up of a 2-cycle together with one or more cycles of odd length we can obtain our 2-cycle by raising this permutation to a suitable power. For example if we found that we could generate (13)(254) we can then generate (13) since

$$(13) = [(13)(254)]^3.$$

As a result of the above observations we have widened our search from looking for a particular 2-cycle to looking for any permutation which is made up from any 2-cycle together with other cycles of odd length. At this stage we must revert to trial and error, but there are so many permutations of this type that it is usually not difficult to find one. Let us try in this case. The permutations t and l on their own are not of this type. Let us combine them. $tl = (13)(48765)$ which is made up from a 2-cycle and a 5-cycle. We have already found what we were looking for. If this first attempt had not proved successful we might have tried tlt, tl^2, $tltl^3$ and various other combinations. With a bit of foresight we would not bother trying lt or t^2l. This is because $lt = t^{-1}(tl)t$ which, being a conjugate of tl, has the same cycle structure as tl, and t^2l in this example is equal to l since $t^2 = I$.

We have discovered that $tl = (13)(48765)$. Thus $(tl)^5 = (13)$. All we need to do now is to conjugate this by some permutation x to obtain (18).

The permutation x must satisfy the following requirements:

(a) under the action of x, $1 \to 1$ and $3 \to 8$,
(b) x must be built up from l and t.

To do this we must examine permutations that can be generated by l and t which fix the symbol 1. One such permutation is $(tl)^2 = (47586)$. Unfortunately this does not send 3 to 8, even if repeated several times because 3 is not in the cycle. Note, however, that t sends 1 to 4 and that l^3 sends 4 to 1. Consequently tl^3 sends 1 to 1. Computing it we find that $tl^3 = (285)(3746)$. Now although 3 and 8 are present in this expression, they are in different cycles. This means that tl^3 can never send 3 to 8, even if repeated several times.

We have discovered two permutations which can be generated by l and t and which fix the symbol 1, viz.

$$(tl)^2 = (47586) \text{ and}$$
$$tl^3 = (285)(3746).$$

Neither, by itself, can do what we require – replace 3 by 8. However a combination of the two can do the job. Firstly note that anything built up from $(tl)^2$ and tl^3 will be built up from l and t. Moreover any such permutation will fix the symbol 1. All we need to do now is to find a permutation built up from $(tl)^2$ and tl^3 which will send 3 to 8. This is easy. For example, $(tl^3)^{-1}$ sends 3 to 6 and $[(tl)^2]^{-1} = (tl)^{-2}$ sends 6 to 8. Thus $(tl^3)^{-1}(tl)^{-2}$ will send 3 to 8. This is a permutation x that will conjugate (13) into (18). Computing x we find that $x = (258)(3647) \cdot (46857) = (2738)$. Thus

$$\begin{aligned}(18) = (13)^{(2738)} &= (2738)^{-1}(13)(2738)\\ &= [(tl^3)^{-1}(tl)^{-2}]^{-1}(tl)^5(tl^3)^{-1}(tl)^{-2}\\ &= (tl)^2(tl^3)(tl)^5 l^{-3} t^{-1}(l^{-1}t^{-1})^2\\ &= (tl)^2 tl^3(tl)^5 l^5 t(l^7 t)^2 \text{ since } l^8 = I \text{ and } t^2 = I.\end{aligned}$$

Note that we do not claim that this is the solution involving the least number of moves. In general there is no simple method for determining the least number of moves in the solution of a permutation problem.

Interpretation of the result

The desired change in the arrangement of the carriages can be achieved by performing the operations l and t according to the pattern

$$\begin{aligned}(tl)^2 tl^3(tl)^5 l^5 t(l^7 t)^2 &= tltltl^3tltltltltll^5tl^7tl^7t\\ &= tltltl^3tltltltltl^6tl^7tl^7t.\end{aligned}$$

Now the permutation *tl* involves operating the turntable on the left-hand four carriages and then taking one carriage from the left-hand end across to the other side so that the turntable can operate on those carriages that were previously in positions 2,3,4 and 5. However, in practice it is not necessary to move the left-hand carriage right across to the other side to give the next four carriages access to the turntable. It is sufficient for the train to move one carriage to the left. Of course if at any time there are insufficient carriages on the turntable, extra ones must then be moved across. Recognizing this, the whole procedure can be given by the following instructions: 1 2 3 [←3] 1 2 3 4 5 3 2 1, where the symbol [←3] indicates that at that stage three carriages must be taken from the right-hand side to the left-hand side, and the remaining digits indicate that the turntable must be operated with the carriage in that position on the left-hand side of the turntable. The following table sets out the successive arrangements of the carriages throughout this procedure (the black rules represent the position of the turntable).

Instructions	Carriages							
	1	2	3	4	5	6	7	8
1								
	4	3	2	1	5	6	7	8
2								
	4	5	1	2	3	6	7	8
3								
	4	5	6	3	2	1	7	8
[←3]								
	1	7	8	4	5	6	3	2
1								
	4	8	7	1	5	6	3	2
2								
	4	5	1	7	8	6	3	2
3								
	4	5	6	8	7	1	3	2
4								
	4	5	6	3	1	7	8	2
5								
	4	5	6	3	2	8	7	1
3								
	4	5	8	2	3	6	7	1
2								
	4	3	2	8	5	6	7	1
1								
	8	2	3	4	5	6	7	1

Jump over the hole

Another permutation puzzle requires seven coins and a grid of nine squares as indicated in the following diagram.

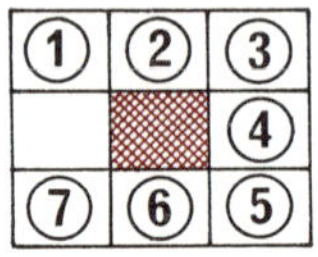

Seven of the eight outside squares are each occupied by a coin and the middle square represents a hole. The coins may move around the edge of the hole, or a coin may 'jump over the hole'. Jumping over the hole must be done in a straight line only, either vertically or horizontally. The object of the puzzle is to produce the following arrangement.

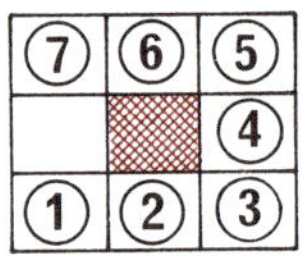

Chessmen's hopscotch

The game of chess was still a long way from finishing. Some of the minor pieces on each side had been captured and were standing around beside the board wondering what they could do to pass the time.

'I cannot understand how you manage to jump about in such an erratic manner,' said the white bishop to one of the knights. 'It looks so difficult to change your direction in mid-air.'

'There's nothing to it,' replied the black knight, 'See, this is how it is done – one along and one diagonally.' And he proceeded to demonstrate the knight's move. He did it so gracefully that it did look easy to the others and they all began to try. Soon they were all hopping around like knights – even the little pawns.

'Would you pieces mind keeping still a while,' said the black queen rather crossly, 'some of us haven't finished the game yet. How can we concentrate on a checkmate with you all jumping about like that? Can't you go over to that other board and play hopscotch or something?'

'What's that?' asked the white bishop, puffing after so much exercise.

'It's a game that we knights play whenever we get a few sets together. I suppose that you other pieces could play now that you know how to jump like we do.'

They went over to an unused board and the black knight marked out eleven squares. He chose one rook, one knight, one bishop and one pawn of each colour and directed them to certain squares.

'Now,' said the black knight, 'at any time anybody may jump into a vacant square. We keep on jumping and we have to see how long it takes us to change ends.'

'That's right,' said the white knight, 'each black piece must end up on the square where his white partner is now standing and vice versa.'

'And must we always jump like a knight?' asked the white bishop who was still a little red in the face from his earlier exercise and who was beginning to think that gliding gently diagonally was much more becoming to his rank.

'I'm afraid so,' said the black knight. 'Those are the rules.'

'And can we capture other pieces?' squeaked a little pawn hopefully.

The black knight laughed. 'Goodness no,' he replied, 'we're off duty now. In this game we co-operate instead of competing. That way nobody loses.'

So they began jumping. But they soon discovered that it was not easy to get everybody into place at one time. So they continued to jump and in fact long after the cry of 'checkmate' had gone up from the other board, they were still jumping.

EXERCISE SET 11

1. Apply the techniques described earlier to solve the following modifications to the shunting problem. However, if a solution does not come easily, do not spend too much time on the problem. At least one of these cases is impossible!

	number of carriages (including engine and guard's van)	*number of carriages to operate turntable*
(a)	6	4
(b)	5	4
(c)	8	3
(d)	7	3

2. Take eight coins and label them with the letters DEARGRAN. Starting with the coins in this order and using the two basic operations of the shunting yard problem, obtain the arrangement ARRANGED.

3. Solve the 'Jump over the hole' puzzle.

4. Solve the 'Chessmen's hopscotch' puzzle.

[HINTS: Number the squares around the edge of the centre square as in the following diagram.

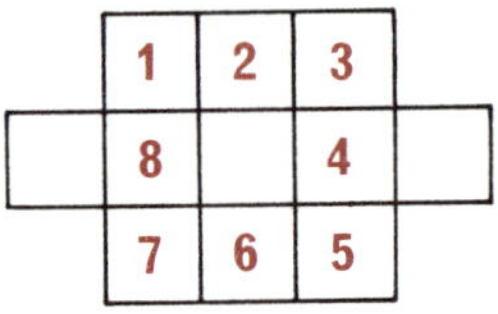

Move the piece in square 6 to the vacant square on the right-hand end. Now move the piece in square 1 to square 6 and continue to move the pieces until square 2 becomes available. Finally move the piece in the end square to square 2. You will find that except where instructed there will be at each stage only one piece you can move (apart from uselessly returning the piece you have just moved). Following these instructions you will therefore produce a definite permutation.

Repeat this operation, but instead of moving the piece in square 1 to square 6, move the piece in square 3. This gives rise to another permutation. Write down these two permutations in cycle notation.

Finally discover how the permutation (26) can be very simply achieved. Use these three permutations to generate the desired permutation, which is (17)(26)(35)(48).]

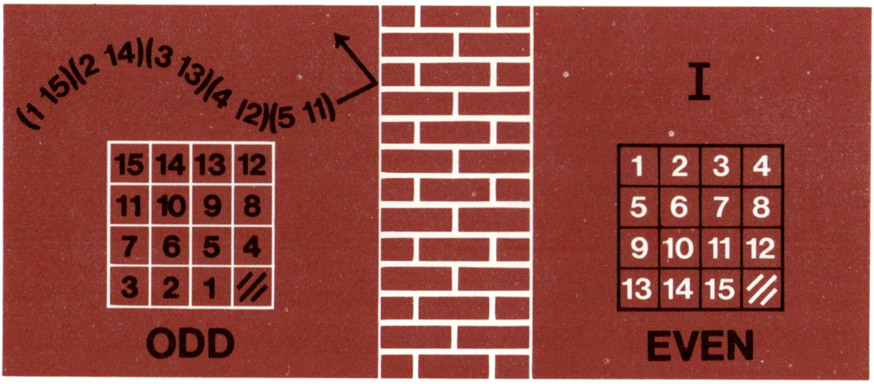

Odd and Even Permutations

Classifying permutations

We saw in chapter 4 that every permutation can be written as a product of 2-cycles. Moreover there are many different ways of breaking a permutation down into 2-cycles. Even the number of 2-cycles is not fixed. For example, (23) = (12)(13)(12). However, it can be shown that the number of 2-cycles must always be odd or it must always be even, depending on the permutation. Thus with some permutations, such as (23), the number of 2-cycles is always odd while with others such as (123) = (12)(13) = (12)(13)(23)(23) etc., the number of 2-cycles must always be even. We use this difference between permutations to classify them as odd or even.

An even permutation is one which can be written as the product of an even number of 2-cycles and an odd permutation is one which can be written as the product of an odd number of 2-cycles.

Since every permutation can be written as a product of 2-cycles, every permutation is either odd or even. And as mentioned above, it can be proved that no permutation is both odd and even.

Examples of odd and even permutations

(124)(35) = (12)(14)(35) and so is odd.
(152)(36978) = (15)(12)(36)(39)(37)(38) and so is even.

(12)(34) is even.
I = (12)(12) and so is even.

Clearly any 2-cycle is odd since it can be written as the 'product' of one 2-cycle. If you object to the product of one permutation you can obtain the same result by attaching (12)(12) to the expression as, for example, (34) = (34)(12)(12).

Every 3-cycle can be written as a product of two 2-cycles (e.g. (123) = (12)(13)) and so every 3-cycle is even. In the same way, since every r-cycle can be written as the product of $r-1$ 2-cycles (e.g. $(123 \ldots r) = (12)(13) \ldots (1r)$), an r-cycle is odd if $r-1$ is odd, that is if r is even, and it is even if r is odd. In other words

a cycle of odd length is an even permutation and a cycle of even length is an odd permutation.

Simple properties of odd and even permutations

The permutation (1234) is odd and (123) is even. The product (1234)(123) = (1342) is odd. This illustrates the fact that the product of an odd permutation with an even permutation is an odd permutation, or simply, 'odd times even is odd'. This result looks strange because it is not true for multiplication of odd and even numbers. However it is quite correct for odd and even permutations. The reason for this is apparent when we observe that when two permutations are multiplied, the number of 2-cycles can be added. (1234)(123) = (12)(13)(14)·(12)(13) where the number of 2-cycles is 3 + 2 = 5. Odd and even permutations behave under multiplication in the same way that odd and even numbers do under *addition*. Thus for permutations,

odd × *odd* = *even*
odd × *even* = *odd*
even × *odd* = *odd*
even × *even* = *even*

We have seen that the inverse of a permutation has the same cycle structure as that permutation. Since the odd-ness or even-ness depends only on the cycle structure, we find that the inverse of an even permutation is even and the inverse of an odd permutation is odd. Another way of reaching this conclusion is to suppose that it were possible for an odd permutation, p, to have an even inverse. Then pp^{-1} must be odd (odd ×

even is odd). But $pp^{-1} = I$ which we agreed is even. This situation can therefore never arise.

Briefly,

$$odd^{-1} = odd$$
$$even^{-1} = even$$

Parity of a permutation

A convenient way of expressing the odd-ness or even-ness of a permutation is to define the *parity* of a permutation. This is a number which is 1 for even permutations and -1 for odd ones. We shall write the parity of a permutation, p, as $P[p]$. For example, $P[(123)] = 1$ since (123) is even, while $P[(12)(34567)] = -1$ since $(12)(34567)$ is odd. The discoveries we made in the previous section can be expressed very nicely in terms of parity.

$$P[\text{even}] = 1$$
$$P[\text{odd}] = -1$$
$$P[\text{I}] = 1$$
$$P[pq] = P[p]P[q] \text{ for all permutations } p, q$$
$$P[p^{-1}] = P[p] \text{ for all permutations } p$$

Parity is the most convenient way to handle problems such as the following.

Example 1. Prove that the square of every permutation, x, is even.

Solution. $P[x^2] = P[x]^2 = 1$ since $1^2 = (-1)^2 = 1$. Hence x^2 is even.

Example 2. Prove that $xyx^2yx^{-1}y$ always has the same parity as y.

Solution. $P[xyx^2yx^{-1}y] = P[x]P[y]P[x^2]P[y]P[x^{-1}]P[y] =$
$P[x]P[y]P[y]P[x]P[y]$ since $P[x^2] = 1$ and $P[x^{-1}] = P[x]$
$= P[x]P[x]P[y]$ since $P[y]^2 = 1$
$= P[y]$ since $P[x]^2 = 1$

Example 3. Prove that if x, y and z are permutations such that $xy = yz$, then zx is even.

Solution. $P[xy] = P[yz]$ and so $P[x]P[y] = P[y]P[z]$ from which we may conclude that $P[x] = P[z]$. Then $P[zx] = P[z]P[x] = P[z]^2 = 1$. Thus zx is even. (Note that we made use of the commutative law in cancelling $P[y]$ from the above equation for the $P[y]$ appears on the left of one expression and on the right of the other. However we are justified in assuming the commutative law for the parities since these are numbers, not permutations. We would *not* be justified in concluding from $xy = yz$ that $x = z$. We could merely conclude that $x = yzy^{-1}$ but without the commutative law there would be no way of the y and y^{-1} 'getting together'.)

It's impossible!

In the last two chapters we considered techniques for generating certain permutations from others. In some cases we found two permutations, x and y, which could generate every permutation on a certain number of symbols (e.g. $x = (12)$, $y = (123 \ldots n)$). However it is possible to find permutations x and y which do not generate all such permutations. A trivial example is $x = \mathrm{I}$, $y = \mathrm{I}$. The only permutation that can be generated by x and y in this case is, of course, I. A less trivial example is $x = (1234)$, $y = (13)(24)$. Observe that $y = x^2$. Thus everything that can be generated by x and y can be generated by x alone. That is, only powers of x can be generated by x and y.

A much less obvious example is $x = (12345)$, $y = (142)$. In this case quite a number of permutations can be generated: 60, in fact. However as there are $5! = 120$ permutations on 5 symbols, this leaves 60 permutations on 1,2,3,4,5 which cannot be generated by x and y. For example we can be quite certain that (12) can never be generated by (12345) and (142). This is because (12345) and (142) are even permutations. Anything which can be built up from them must be even (even times even is even). However (12) is odd. No odd permutation can possibly be generated by two even permutations. (This does not prove that every even permutation can be generated, though this is true for (12345) and (142).)

The concept of odd and even permutations is a valuable one in studying permutations since it shows what *cannot* be done. Perhaps this sounds very negative, but it is just as important in mathematics to know what cannot be done as to know what can. Proving something impossible saves a lot of wasted time in trying to do it. In fact permutations were first introduced into mathematics in order to prove something impossible. For centuries mathematicians had sought to find a formula for the solutions to the general polynomial equation of the fifth degree (the general quintic): $ax^5 + bx^4 + cx^3 + dx^2 + ex + f = 0$, whereby the solutions can be calculated from the coefficients using only the operations of addition, subtraction, multiplication, division and extraction of roots. The reader is probably familiar with the formula for the solutions to the quadratic equation (degree two): $ax^2 + bx + c = 0$, viz.

$$x = \frac{-b \pm \sqrt{b^2 - 4ac}}{2a}$$

Formulae of the same type (though much more complicated, and involving cube roots) exist for polynomial equations of degree three and four. These were known to mathematicians for about two centuries before the birth of Galois and there seemed no reason to believe that a formula would not exist for degree five. The fact that one had not been discovered was thought to indicate merely that it was exceedingly complicated. Perhaps if they tried a little harder . . . However the two great mathematicians Abel and Galois put an end to all this vain effort by showing that there was no such formula, not only for degree five, but for all higher degrees as well. To this day this branch of mathematics is known as Galois Theory. This theory depends on much deeper permutation theory than that of odd and even permutations and is beyond the scope of this book. We shall content ourselves with discovering impossibilities of a much more pedestrian kind.

The four-square puzzle

An interesting application of odd and even permutations is to a puzzle that consists of a plastic frame containing 15 small plastic tiles numbered from 1 to 15. They are arranged in the frame in four rows of four with one empty space. The tiles can be moved around in the frame vertically or horizontally by moving a tile into the empty space. The problem is to obtain specified patterns. One pattern that is usually given is labelled 'IMPOSSIBLE'. Using the theory of permutations we shall see why this pattern is impossible.

1	2	3	4
5	6	7	8
9	10	11	12
13	14	15	16

Fig. 1

15	14	13	12
11	10	9	8
7	6	5	4
3	2	1	16

Fig. 2

We number the positions 1 to 16 as in Fig. 1. The tiles are numbered from 1 to 15 and we number the empty space 16. The puzzle is usually set up with tiles occupying their corresponding positions and with 'tile' 16 (the empty space) occupying position 16 in the bottom right-hand corner.

Everything that can be done with the puzzle can be built up from the following basic moves: interchange tile 16 (the empty space) with the tile above, the tile below, the tile on the left, or the tile on the right. The permutations they correspond to are 2-cycles. For example, with the puzzle set up as in Fig. 1, the first permutation can be either (15 16) or (12 16). The important thing to remember is that these basic permutations are all *odd.*

Now imagine a sequence of these basic permutations being carried out so that 'tile' 16 reverts to its original place. This must involve an even number of vertical moves (every move up must eventually be balanced by a move down and conversely) and an even number of horizontal moves. In all, an even number of these basic moves must be made. The permutation that corresponds to the whole sequence of moves is therefore the product of an even number of odd permutations. That is, it is an even permutation.

Thus if the empty space is to return to the position from which it came, only *even* permutations are possible. If the puzzle is set up as in Fig. 1, then the arrangement given in Fig. 2 is impossible, for the permutation involved in going from one to the other is (1 15)(2 14)(3 13)(4 12)(5 11)(6 10)(7 9) which is an *odd* permutation.

How many even permutations are there?

The $n!$ permutations on n symbols can be put into two classes – the odd ones and the even ones. Which is the bigger class? Or are there just as many odd permutations as even ones?

Firstly there is the very trivial case where $n = 1$. The only permutation on one symbol is I and this is even. So for $n = 1$, there is one even permutation and no odd ones. But this is an exceptional case. For all larger values of n there is the same number of each, as we shall see. Since there are $n!$ permutations altogether, this will lead to the conclusion that there are $\frac{1}{2}n!$ of each.

Suppose that the number of even permutations is r and the number of odd permutations is s. If we multiply the r even permutations on the left by (12), they will produce r different permutations (if $(12)x = (12)y$

then multiplying on the left again by (12) we must have $x = y$). Moreover, these r permutations will be odd. Thus there are at least r different odd permutations and so $s \geqslant r$. Starting with the s odd permutations and multiplying them on the left by (12) we produce s different even permutations. Thus $r \geqslant s$. Since $s \geqslant r$ and $r \geqslant s$, it must be that $r = s$. The number of even permutations is equal to the number of odd ones.

Note that in the above argument we needed to have $n \geqslant 2$ in order to be able to use the 2-cycle (12). It is just as well that the argument breaks down for $n = 1$ because, as we saw, the result is false in that case!

The argument gives a way of pairing off the even permutations with the odd ones: p is paired off with $(12)p$. The following table lists the 24 permutations on 1,2,3,4 and shows how the 12 even permutations can be paired off with the 12 odd ones.

S_4

even	*odd*
I	(12)
(123)	(13)
(132)	(23)
(124)	(14)
(142)	(24)
(134)	(1234)
(143)	(1243)
(234)	(1342)
(243)	(1432)
(12)(34)	(34)
(13)(24)	(1423)
(14)(23)	(1324)

EXERCISE SET 12

1. Which of the following are even permutations?
   (a) (123456789)
   (b) (123456)(789)
   (c) (123)(4567)(89)
   (d) (12)(345)(67)(89)
   (e) (123)(456)(789)

2. Which of the following statements are true?

(a) The inverse of an even permutation is odd.

(b) The product of two odd permutations is also odd.

(c) $P[xy] = P[yx]$ for all permutations x, y.

(d) $P[x^{-1}yx] = P[y]$ for all permutations x,y.

(e) If $n > 1$, there are $\frac{1}{2}n!$ even permutations on $1,2,3, \ldots, n$.

3. Prove for all permutations x, y, that $x^{-1}y^{-1}xy$ is an even permutation.

4. Examine any case in question 1 of exercise set 11 for which you were unable to find a solution. Can you now prove that the problem is impossible in such a case?

Groups of Permutations

What is a group of permutations?

The product of two even permutations is even. That is, we can never get outside the set of even permutations by multiplying even permutations. We describe this phenomenon by saying that the set of even permutations is *closed.* Any set of permutations is defined to be closed if the product of any two permutations in the set is again in the set.

We have already met two examples of closed sets of permutations on $1,2,3,\ldots,n$, viz. the set of all permutations which we denoted by S_n, and the set of all even permutations (remember that even × even = even) which we denote by A_n. An example of a set of permutations which is not closed is $\{I, (1234), (13)(24)\}$ since $(1234)\cdot(13)(24) = (1432)$ which is not in the set. However if we include (1432) we do obtain a closed set, viz. $\{I, (1234), (13)(24), (1432)\}$.

The question raised by the title of this section is now easily answered. A *group* of permutations is simply a closed set of permutations (provided that the set contains at least one permutation). For example, S_4, A_4, $\{I, (1234), (13)(24), (1432)\}$ are all groups. Moreover all of these groups are subsets of S_4, that is, their elements are elements of S_4. We describe this by saying that they are *subgroups* of S_4. Any group which is contained in a group is said to be a subgroup of it.

What must be in a group?

Suppose that a certain group contains the permutation, x. Then the product x^2 (= $x \cdot x$) must also be in the group. Furthermore, x^3 (= $x^2 \cdot x$), x^4 (= $x^3 \cdot x$) etc. must all be in the group. Every positive power of x, in fact, must be in the group. Now we have seen that eventually some power of x, say x^n, is equal to the identity permutation, I. So taking x to be any permutation in the group we can see that the group must also contain I.

The inverse of a permutation x, can be written as a positive power of x. For if $x^n = \mathrm{I}$, then $x^{n-1} = x^{-1}$. So if a group contains x, it must therefore contain x^{-1}. And in fact it must contain all of the positive powers of x^{-1}, that is, all of the negative powers of x. We have also seen that it contains x^0 (this is simply I). Thus

> *a group which contains x contains every power of x (including x^{-1} and* I).

Every group must contain I. There is in fact a group that contains I and nothing else. For the set $\{\mathrm{I}\}$ is closed.

Multiplication tables

The basic method for testing a set of permutations to see if it is closed is to test all possible products. The easiest way to do this is to set up a multiplication table. If at any stage there appears a permutation which is outside the set you need go no further – the set is not closed. If on the other hand the table can be completed with every product being one of the permutations in the set, the set is closed – it is a group.

Let us carry out this procedure on the set $\{$I, (12)(34), (13)(24)$\}$. We follow the usual convention that the first factor is found in the left-hand column and the second factor is found in the top row.

	I	(12)(34)	(13)(24)
I	I	(12)(34)	(13)(24)
(12)(34)	(12)(34)	I	(14)(23)
(13)(24)			

We need go no further as (14)(23) is not in the set. If we include it we do obtain a group. Draw up the multiplication table for $\{$I, (12)(34), (13)(24), (14)(23)$\}$ and check for yourself that it is a group. This group is usually given the symbol V_4.

EXERCISE SET 13

1. Which of the following sets of permutations are groups?
   (a) $\{(123), (132)\}$
   (b) $\{I, (1234), (1243), (1342), (1324), (1423), (1432)\}$
   (c) $\{I, (12), (34), (12)(34)\}$
   (d) $\{(12)\}$
   (e) $\{I, (13), (24), (12)(34), (13)(24), (14)(23), (1234), (1432)\}$

2. Write out the multiplication table for the set $\{I, (123), (12)(34)\}$. Every time you obtain a permutation which is not in the set, include it in the set and extend the multiplication table. Continue until you have enlarged the set just enough to make it closed. Which group do you obtain?

Cyclic groups

If a group contains a permutation, x, it must contain all of the powers of x. It is possible, however, for it to contain nothing else. Some groups consist of a single element, x, together with its powers, that is, they are generated by one element, x.. Such groups are called *cyclic* groups.

For example, $\{I, (1234), (13)(24), (1432)\}$ is a cyclic group generated by $x = (1234)$ since all of the permutations in the group are powers of x.

$$\begin{aligned} I &= x^0 \\ (1234) &= x^1 \\ (13)(24) &= x^2 \\ (1432) &= x^3 \end{aligned}$$

In fact this group can also be generated by $y = (1432)$ since

$$\begin{aligned} I &= y^0 \\ (1234) &= y^3 \\ (13)(24) &= y^2 \\ (1432) &= y^1 \end{aligned}$$

This group cannot be generated by $(13)(24)$, for the group generated by $(13)(24)$ is $\{I, (13)(24)\}$.

The number of elements in a cyclic group is equal to the order of its generator. For if x has order n, the first time that the sequence of powers $x^0, x^1, x^2, \ldots$ repeats is at the $(n + 1)$th place: $x^n = x^0 = I$. This means that there are precisely n different powers of x and so n permutations in the cyclic group generated by x.

Conversely if a group contains n permutations and one of these has order n then the group must be cyclic because the n different powers of that permutation must fill up the whole group.

This little fact is useful in testing to see if a group is cyclic. Count the number of elements in the group. If the group contains a permutation whose order is equal to this number, the group is cyclic. Otherwise the group is not cyclic.

For example, $V_4 = \{I, (12)(34), (13)(24), (14)(23)\}$ is not cyclic since it contains 4 permutations yet none of them has order 4. On the other hand $\{I, (123), (132)\}$ is a cyclic group since it contains 3 permutations including two of order 3. Either of these can be used to generate the group.

Every element in a group generates a cyclic subgroup. By considering each element in turn we can find all of the cyclic subgroups of the group. For example, the following are the cyclic subgroups of S_3.

generator	*cyclic subgroup*
I	$\{I\}$
(12)	$\{I, (12)\}$
(13)	$\{I, (13)\}$
(23)	$\{I, (23)\}$
(123)	$\{I, (123), (132)\}$
(132)	$\{I, (132), (123)\}$

Note that the last two cyclic subgroups are the same. (The order in which the permutations are listed is immaterial.) It is usual for repetitions such as this to occur because a cyclic subgroup containing 3 or more permutations is generated by more than one permutation.

Package-preserving permutations

Mr and Mrs Jones and Mr and Mrs Smith went on holiday together by car. As they could all drive they decided to take turns in driving. Rather than take turns in strict rotation they thought it might be more interesting if they used permutations to decide the change in seating arrangements after every hundred miles. Mr Jones prepared 24 cards and on each card he wrote one of the 24 permutations on 1,2,3,4. They numbered the seats 1,2,3,4 as in the diagram and set off with Mr and Mrs Jones and Mr and Mrs Smith occupying seats 1,2,3 and 4 in that order. After the first hundred miles they happened to choose the permutation (124) from the pack. This meant that Mr Jones who had been sitting in seat number 1 had to move to seat number 2 and Mrs Jones had to move from seat number 2 to seat number 4. Mr Smith who had been sitting in seat number 3 stayed where he was and Mrs Smith went from seat

number 4 to seat number 1, next to Mr Jones. This meant that Mr Jones and Mrs Smith were sitting together in the front and Mr Smith and Mrs Jones were in the back. They decided that this was not very satisfactory as Mrs Jones and Mrs Smith wanted to sit next to their own husbands. So they decided to discard from the pack all those permutations which would split up the two couples and keep only the 'couple-preserving' permutations.

They found that there were 8 couple-preserving permutations: I, (12), (34), (12)(34), (13)(24), (14)(23), (1324), (1423). For example, if at the first change-over point they had chosen the permutation (1423), Mr Jones would have gone from seat number 1 to seat number 4 and Mrs Jones would have gone from seat number 2 to seat number 3. As a result they would both have gone from the front of the car to the back and so they would still be sitting together.

The important fact about 'couple-preserving' permutations is that if one such permutation is followed by another, the overall effect would be for the couples to still be sitting together. That is, the product of two 'couple-preserving' permutations is again a 'couple-preserving' permutation. Hence these 8 'couple-preserving' permutations form a group.

A similar argument holds in general if we package the symbols which we are considering in some way. There is no need for the 'packages' to be couples. Moreover, the packages need not all be of the same size. We might, for example, take the symbols 1,2,3,4,5,6,7,8,9 and package them as follows: $\{1,5\}$, $\{2,9\}$, $\{3,6,7\}$, $\{4\}$, $\{8\}$. The set of all those permutations which keep these packages intact, is a group. Packaging is a handy way of discovering groups, although not every group of permutations can be found in this way. The following table gives some examples of subgroups of S_4 which can be found by packaging.

packages	*group*
{1}, {2}, {3}, {4}	S_4
{1,2}, {3,4}	{I, (12), (34), (12)(34), (13)(24), (14)(23), (1324), (1423)}
{1,3}, {2,4}	{I, (13), (24), (13)(24), (12)(34), (14)(23), (1234), (1432)}
{1,2,3}, {4}	{I, (12), (13), (123), (132)} = S_3

EXERCISE SET 14

1. Which of the following groups are cyclic?
   (a) {I, (12), (34), (12)(34)}
   (b) {I, (1234), (13)(24), (1432)}
   (c) {I, (12)(34), (13)(24), (14)(23)}
   (d) {I, (1234)(56), (13)(24), (1432)(56)}
2. Is {I, (12345), (15432), (13254), (14523)} a cyclic group?
3. Write down the cyclic groups generated by the following permutations:
   (a) (134); (b) (134)(56)
4. Write down the group of permutations which leave the following packages intact: {1,2,3}, {4,5}.
5. Write down the 24 *even* permutations which leave the following packages intact: {1,2}, {3,4}, {5,6}.

Groups of symmetries

If the corners of a square are numbered 1,2,3 and 4 and the square is rotated or turned over (or both) the corners will undergo some permutation. For example, rotating the square through 90 degrees clockwise corresponds to the permutation (1234).

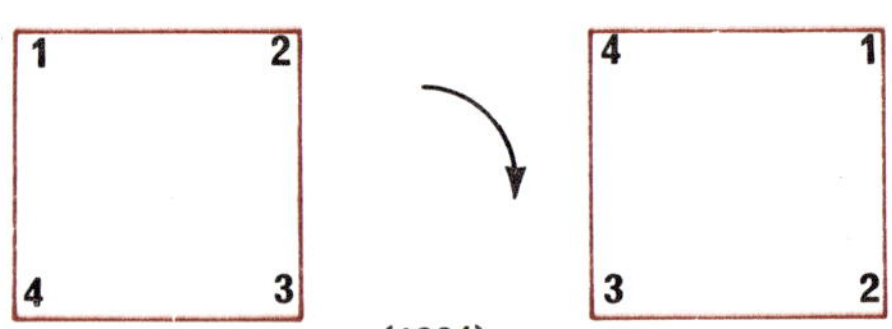

(1234)

Turning the square over about the horizontal axis corresponds to the permutation (14)(23).

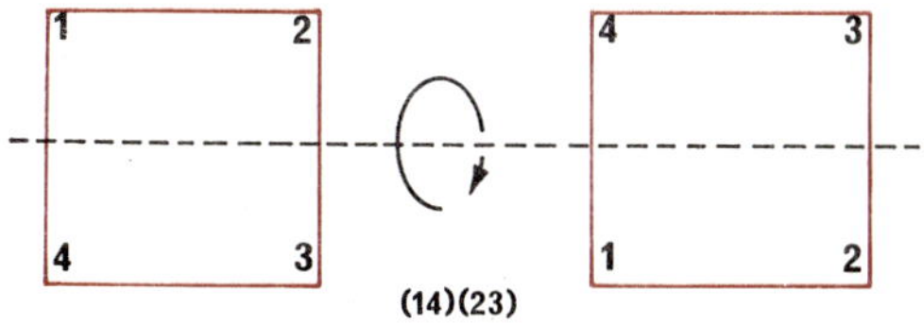

(14)(23)

Turning the square over about the bottom left - top right diagonal corresponds to the permutation (13).

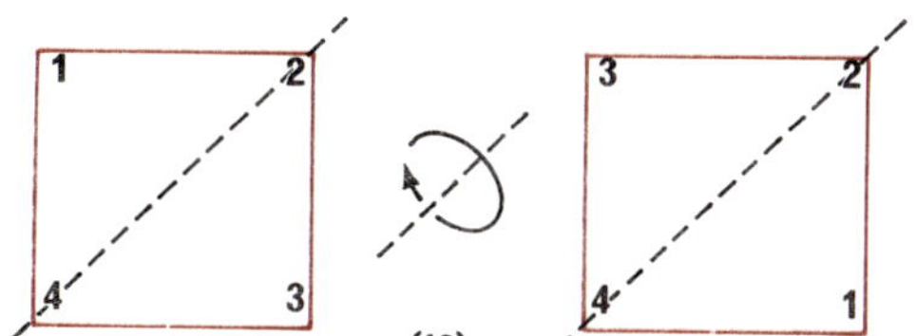

(13)

Not every permutation on 1,2,3,4 can be obtained in this way. For example, (12) cannot. For it is impossible, without twisting the square, to interchange corners 1 and 2 while keeping corners 3 and 4 fixed. However, the product of two permutations which can be obtained by rotating the square can itself be so obtained. We simply carry out the two rotations one after the other. For example, (1234) × (14)(23) = (13). The permutation (13) can be obtained by first rotating the square through 90 degrees clockwise and then rotating through 180 degrees about the horizontal axis. In fact, as we have seen, (13) can be achieved by a single rotation.

The product of two permutations which arise from rotations of a square is itself one of these permutations and so the set of all such permutations is a group. It consists of the 8 permutations I, (1234), (1432), (12)(34), (13)(24), (14)(23), (13), (24). This group is called the *group of rotations* of a square.

Now suppose that instead of a square, the figure is a rectangle with unequal sides. The amount of symmetry is reduced and this cuts down the number of permissible permutations. For example, (13) can no longer be obtained since the operation of rotating the rectangle about the bottom left-top right diagonal does not permute the corners. It changes the overall configuration.

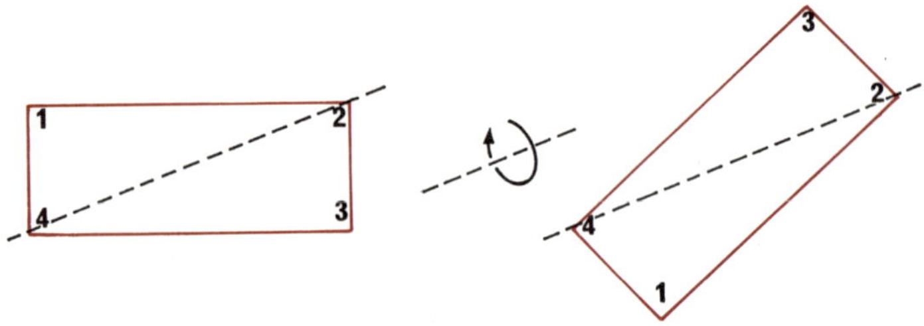

The group of rotations of a rectangle is $V_4 = \{I, (12)(34), (13)(24), (14)(23)\}$.

If the figure is merely a typical parallelogram the group of rotations is cut back further to $\{I, (13)(24)\}$.

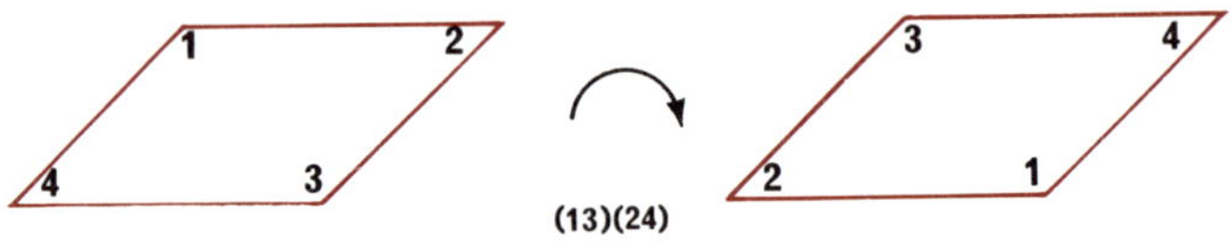

Finally, if the figure is a quadrilateral with no symmetry at all, the group of rotations consists of the identity permutation alone.

Three-dimensional figures have groups of rotations also. The group of rotations of a regular tetrahedron is A_4. For example, (12)(34) corresponds to a rotation through 180 degrees about an axis joining the midpoint of corners 1 and 2 to the midpoint of corners 3 and 4.

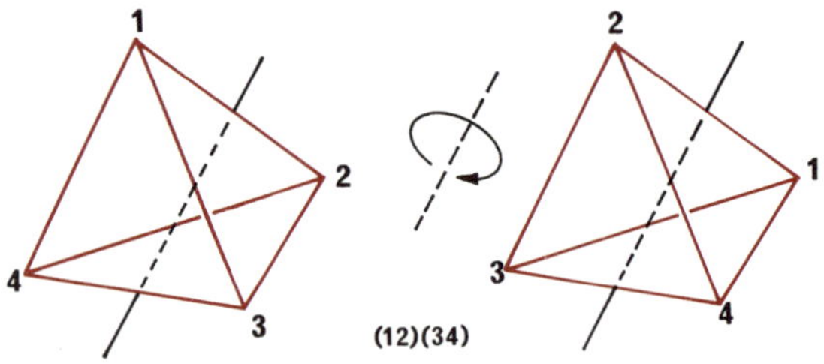

The permutation (123) corresponds to a rotation through 120 degrees about an axis joining corner 4 to the centre of the opposite face.

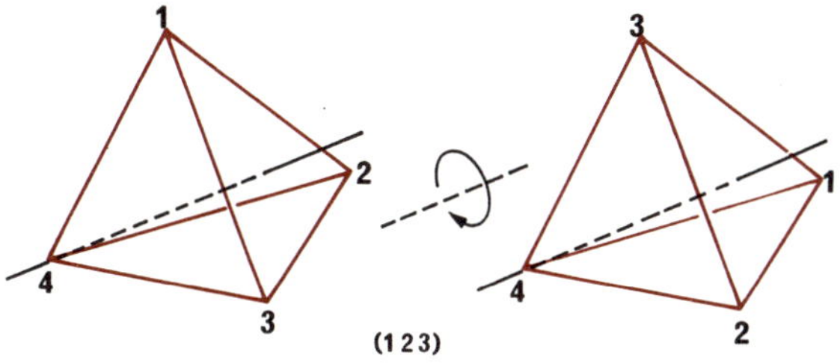

Instead of numbering the four corners we could have achieved the same result by numbering the four faces. However the group of rotations would be the same.

In the case of a cube, it is simpler to number the faces than the corners as there are only 6 faces, but 8 corners.

No matter how we rotate a cube, opposite faces will always remain opposite. We can therefore regard the three pairs of opposite sides as three packages. Only permutations which keep these three packages intact can be obtained by rotating a cube. Furthermore it can be shown that only even permutations can be so obtained. The group of rotations of a cube is the group of all even permutations on 1,2,3,4,5,6 which leave intact the packages $\{1,2\}$, $\{3,4\}$, $\{5,6\}$ (where the faces are numbered so that 1 and 2 are opposite etc.). This is the group you were asked to find in question 5 of exercise set 14.

Another version of the group of rotations of a cube is obtained if we number the vertices, but number opposite corners with the same number.

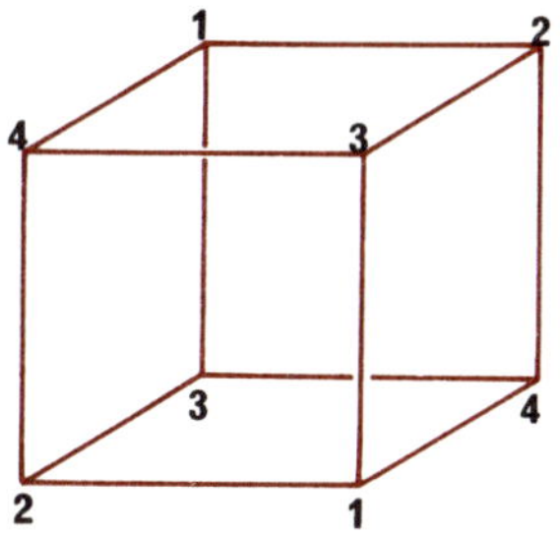

The group of rotations is now S_4. Every permutation on 1,2,3,4 can be obtained. The following diagrams illustrate some of the possible rotations and the permutations they correspond to.

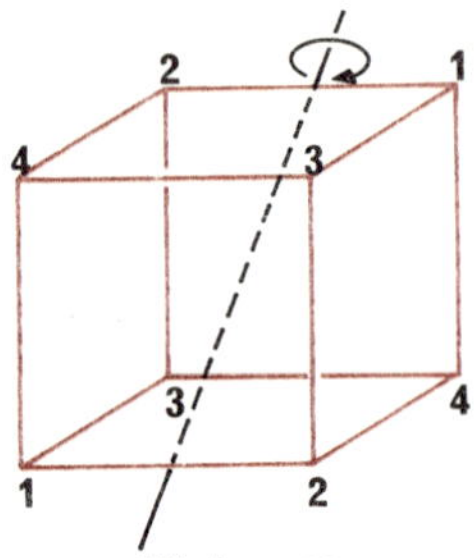

180 degrees
(12)

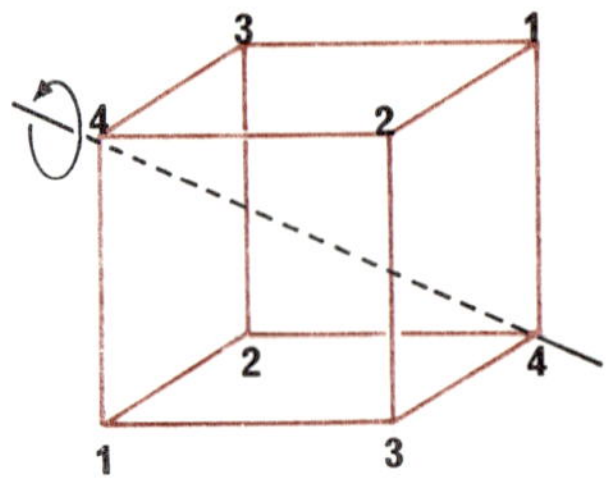

120 degrees
(123)

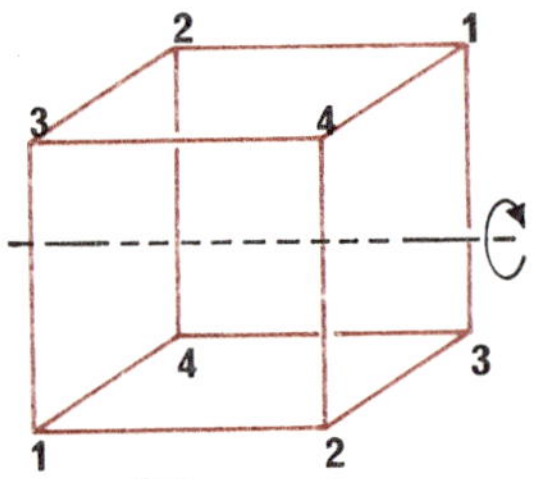

180 degrees
(12)(34)

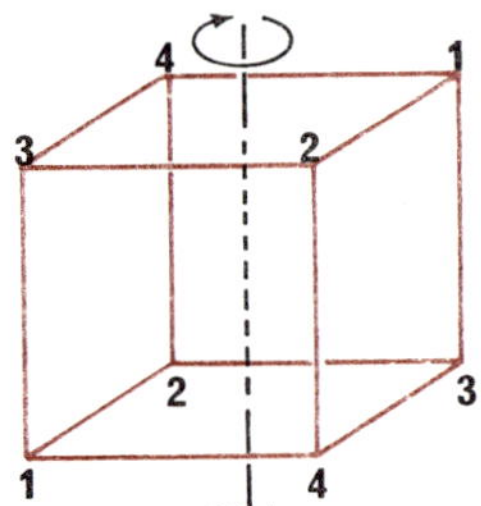

90 degrees
(1234)

Note that the angle through which the rotation takes place is connected with the order of the permutation. A rotation through 90 degrees when carried out 4 times gives a rotation through 360 degrees, which corresponds to the identity permutation. The original permutation therefore has order 4. A permutation of order 3 corresponds to a rotation through 120 degrees, and so on.

In how many ways can the six faces of a cube be coloured using six different colours (each face a different colour)? If we numbered the faces from 1 to 6 there would be 6! = 720 different ways of colouring the cube, since every arrangement of the colours would give a way of associating a colour with each face. However the faces are not numbered. Before painting them, the faces are identical. Many of these 720 'different' ways of colouring the cube will in fact be the same. The two cubes below are identically coloured – each can be obtained from the other by a suitable rotation.

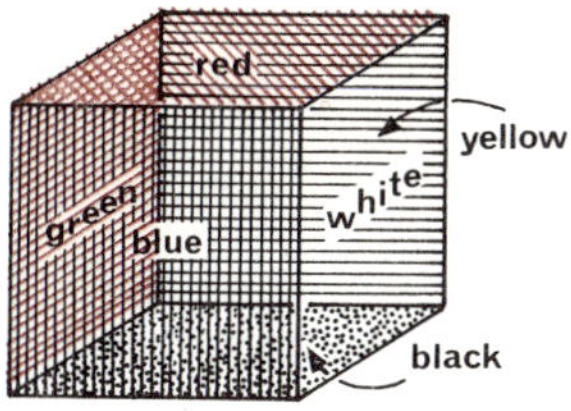

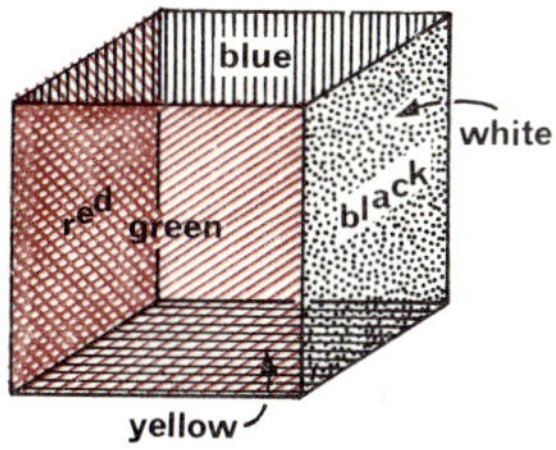

The group of rotations of a cube contains 24 permutations. This means that every way of colouring the cube will be included 24 times in the above count of 720. The number of different ways of colouring a cube is therefore 720/24 = 30.

The number of different ways of colouring an object having n faces with n different colours is $n!/s$ where s is the number of permutations in the group of rotations of the object. This can be deduced using a similar argument to the above. For example, the number of ways of colouring the four sides of a regular tetrahedron is $4!/12 = 24/12 = 2$, since the group of rotations of the regular tetrahedron is A_4 which contains 12 permutations. These two different ways of colouring a regular tetrahedron are illustrated below.

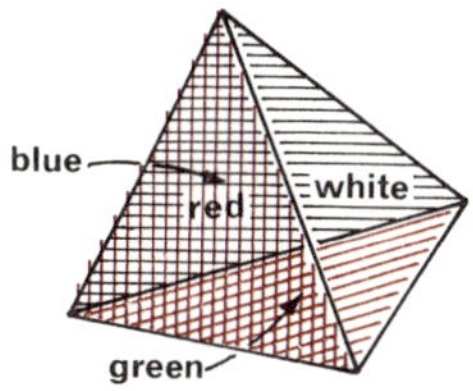

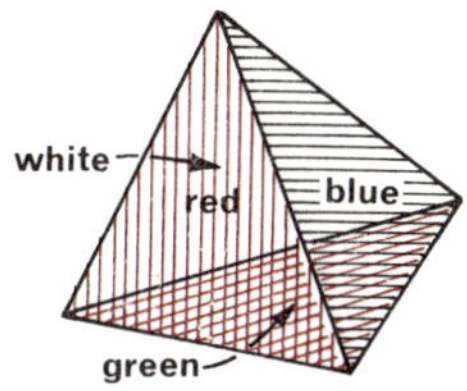

EXERCISE SET 15

1. Find the group of rotations of an equilateral triangle.

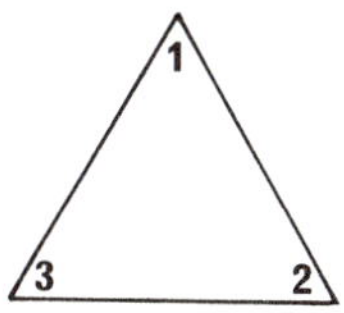

2. Find the group of rotations of a rhombus.

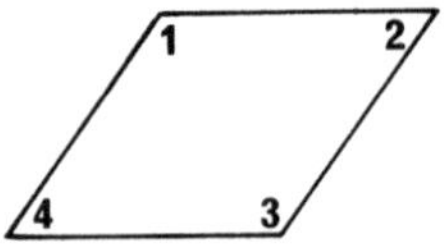

3. Find the number of permutations in the groups of rotations of the following figures.

(a)

(b)

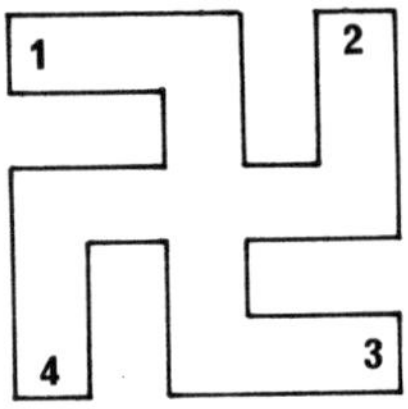

(c)

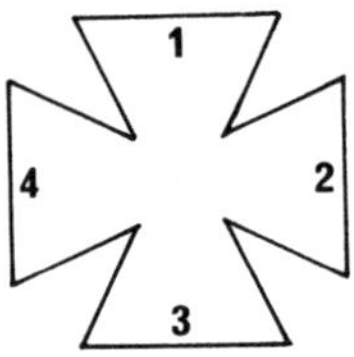

(d)

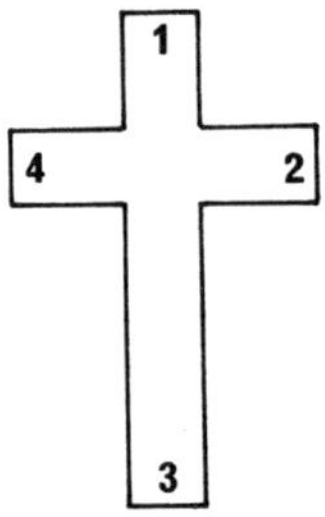

4. Write down the group of rotations of the following rectangular box with square ends. In how many ways can these six faces be coloured using six different colours?

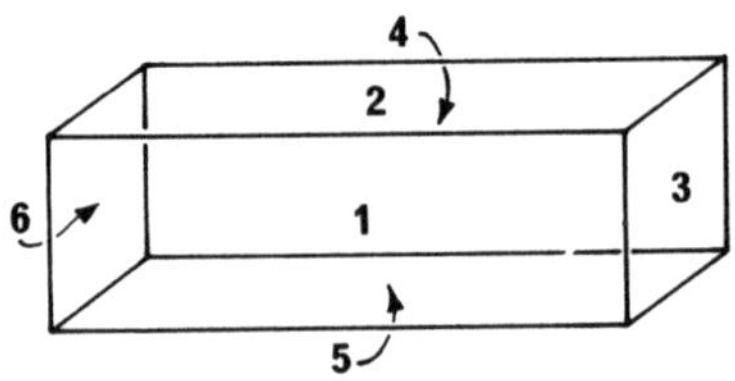

5. (Hard.) Cut out, from a sheet of cardboard, 12 pentagons with each side measuring 2 in (or 5 cm). Join these 12 faces so that they form a closed 3-dimensional object. This is called a dodecahedron. Number the faces from 1 to 12. How many permutations are there in the group of rotations of a dodecahedron? In how many ways can a dodecahedron be coloured using 12 colours?

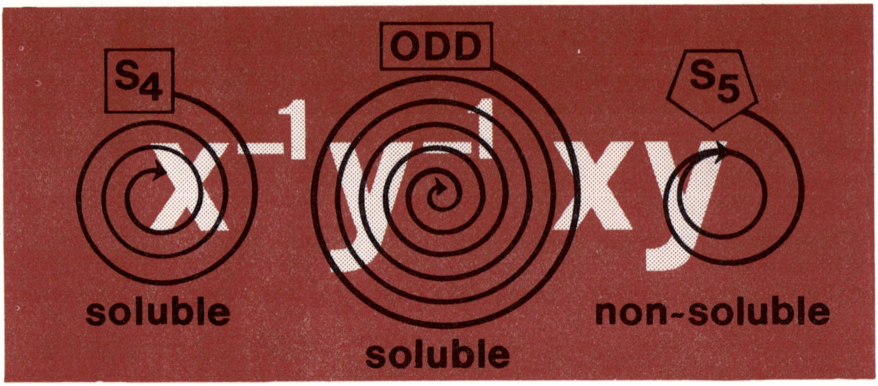

Abstract Groups

What is a group?

Although group theory began with groups of permutations, the subject today includes groups of many things other than permutations – groups of numbers, groups of functions and so on. We know what a group of permutations is. What is meant just by the word 'group'?

Firstly, a group is a collection of objects, or to use the more mathematical terminology, a set of elements. The elements might be permutations, they might be numbers or they might be matrices (see volume 19 of this series). As far as group theory is concerned, it does not matter what these elements are. What *is* important is that the elements combine together in some way. For convenience we usually refer to this combination as multiplication and the result of multiplying two elements as their product. These were the words we used for permutations. But although we borrow the words 'multiplication' and 'product' from arithmetic, we use them merely by way of analogy. The elements need not be numbers or permutations, and even if they are numbers, the method of combination need not necessarily be ordinary arithmetical multiplication.

A group is a set of elements which combine together in some way. A further requirement is that the group must be closed. The product of any two elements must again be an element of the group.

There are three further requirements for a group. We did not mention them when we spoke of groups of permutations because they are automatically fulfilled in that case. For the general case we must write them down explicitly.

The first of these is the associative law:

$$x(yz) = (xy)z \text{ for all elements } x, y \text{ and } z \text{ in the group.}$$

We saw that multiplication of permutations is associative. But not every method of combining elements is associative. An example to show this can be found in ordinary arithmetic. Subtraction is not associative as, for example, $3 - (2-1) = 2$, while $(3-2)-1 = 0$.

The second requirement is that the group must contain an identity element. We borrow the symbol, 1, from arithmetic to denote the identity of a group. It has the property that

$$1x = x1 = x \text{ for all elements, } x, \text{ in the group.}$$

In the case of permutations this identity element is, of course, the identity permutation, I.

The last requirement is that every element of the group must have an inverse. Again we borrow the notation from arithmetic. The inverse of the element, x, is denoted by x^{-1}. The inverse, x^{-1}, has the property that $xx^{-1} = x^{-1}x = 1$.

A group, G, *is a set of elements which combine together in some way such that*

(a) the product of any two elements is in the group
(b) the associative law holds
(c) the group has an identity element
(d) every element has an inverse

Examples of groups

(1) *Groups of permutations.*

(2) *Group of non-zero real numbers.* The set of all non-zero real numbers is a group if the group multiplication is taken to be ordinary arithmetical multiplication. The product of any two non-zero real numbers is a non-zero real number. Multiplication of real numbers is associative. The identity is the number 1. The inverse of any non-zero real number, x, is the inverse $1/x$ or x^{-1} in the usual sense.

(3) *Group of real numbers.* The set of all real numbers is a group if we take the group multiplication to be ordinary arithmetical addition. The identity is the number 0 and the inverse of x is $-x$. The multiplicative notation that is used for groups is a little confusing here. However if you translate the definition of a group by replacing 'product' by 'sum', 'xy' by '$x + y$', '1' by '0' and 'x^{-1}' by '$-x$', you will find that all four requirements are fulfilled in this case.

(4) *Group of quaternions.* The group of quaternions consists of the following eight elements: 1, −1, i, $-i$, j, $-j$, k, $-k$. They combine according to the following multiplication table.

	1	−1	i	$-i$	j	$-j$	k	$-k$
1	1	−1	i	$-i$	j	$-j$	k	$-k$
−1	−1	1	$-i$	i	$-j$	j	$-k$	k
i	i	$-i$	−1	1	k	$-k$	$-j$	j
$-i$	$-i$	i	1	−1	$-k$	k	j	$-j$
j	j	$-j$	$-k$	k	−1	1	i	$-i$
$-j$	$-j$	j	k	$-k$	1	−1	$-i$	i
k	k	$-k$	j	$-j$	$-i$	i	−1	1
$-k$	$-k$	k	$-j$	j	i	$-i$	1	−1

It is associative. For example, $(-i)[j(-1)] = (-i)[-j] = k$ and $[(-i)j](-1) = [-k](-1) = k$. The identity is 1 and the inverses of $1,i,j,k$ are -1, $-i$, $-j$, $-k$ respectively.

(5) *Groups of functions.* Consider the following four functions on the set of non-zero real numbers:

$$\begin{aligned} f(x) &= x \\ g(x) &= -x \\ h(x) &= 1/x \\ k(x) &= -1/x \end{aligned}$$

These form a group if we define the product of two functions in the same way that we defined the product of two permutations. (In fact these functions are permutations on the infinite set of non-zero real numbers, although we cannot write them in cycle notation.) To form the product of two functions, simply follow one by the other. For example,

$$x \xrightarrow{g(x)} -x \xrightarrow{k(x)} -(1/-x) = 1/x$$

and so $gk = h$. The following multiplication table shows how these functions combine with one another.

	f	g	h	k
f	f	g	h	k
g	g	f	k	h
h	h	k	f	g
k	k	h	g	f

The identity of this group is f and every function is its own inverse.

Identities and inverses

Every group has an identity element, 1. Is it possible for a group to have more than one identity element? If so, we had better refrain from referring to *the* identity element. Fortunately there is no problem. We can show that every group has only one identity. For if $1'$ and $1''$ are identity elements for some group then $1'1'' = 1''$ since $1'$ is an identity, and $1'$ because $1''$ is an identity. Therefore $1' = 1''$.

Every element of a group has an inverse. Is it possible for an element to have more than one inverse? Again, the answer is 'no'. To prove this we suppose that x has inverses y and z. Then

$$xy = yx = 1 = xz = zx.$$

By the associative law, $y(xz) = (yx)z$. Using the above equations, this becomes $y = z$.

Soluble groups

The commutative law,

'$xy = yx$ for all elements x and y',

is not necessary for a group. Some groups are commutative (such as the group of non-zero real numbers under ordinary multiplication) and others are not (such as S_4). If the commutative law does hold for a group we say that the group is Abelian* (or commutative).

Typical elements of the cyclic group generated by x are x^m and x^n and since

$$x^m x^n = x^{m+n} = x^n x^m$$

* The adjective 'Abelian' is usually capitalized because it is used in honour of the mathematician Abel.

all cyclic groups are Abelian. Not every Abelian group is cyclic, however. For example, V_4 is Abelian but is not cyclic.

If $xy = yx$, then multiplying both sides of this equation by $x^{-1}y^{-1}$ we obtain

$$x^{-1}y^{-1}xy = yxx^{-1}y^{-1} = 1.$$

An element of a group which can be written in the form $x^{-1}y^{-1}xy$ is called a *commutator.* For example, (142) is a commutator of S_4 since $(142) = (12)^{-1}(143)^{-1}(12)(143)$.

In an Abelian group the only commutator is 1 but in a non-Abelian group there are other commutators. The smallest subgroup containing the commutators is called the commutator subgroup of the group. The commutator subgroup of an Abelian group is simply $\{1\}$.

Some examples of commutator subgroups are given in the table below.

group	*commutator subgroup*
S_4	A_4
A_4	V_4
V_4	$\{1\}$
S_5	A_5
A_5	A_5

Notice what has been done in the first three lines of this table. We started with S_4. The commutator subgroup of S_4 is A_4. We then asked for the commutator subgroup of that. It turns out to be V_4. The commutator subgroup of V_4 is $\{1\}$. Thus the commutator subgroup of the commutator subgroup of the commutator subgroup of S_4 is the subgroup which contains just the identity element.

If, when we carry out this process, we ultimately reach $\{1\}$, we say that the group is *soluble*. Every Abelian group is soluble because we reach $\{1\}$ after just one step. And S_4, though not Abelian, is soluble.

The surprising thing is that when we try this with S_5, instead of 'getting down to the bottom' we 'get stuck' on A_5. The process begins in the same way. The commutator subgroup of S_5 is A_5. But the commutator subgroup of A_5 is A_5 itself. There is no subgroup analogous to V_4. S_5 is therefore *not* soluble.

This difference lies at the heart of the birth of group theory. As explained in chapter 6, group theory began in connection with the problem of *solving* a polynomial equation – that is, of finding a formula for the solutions. Galois showed how a group can be associated with every such equation, called the Galois group of the equation, and he proved that the equation can be solved if, and only if, its Galois group is soluble. This is how the term 'soluble' came to be applied to groups.

Galois also proved that the Galois group of the general polynomial of degree n is S_n. So, it is because S_5 is not soluble that there is no formula for the solutions to a general quintic equation. S_4 is soluble and so the general quartic equation is soluble. Likewise, S_3 and S_2 are soluble and that is why there are formulae for the solutions to the cubic and quadrati equations.

Sizes of subgroups

The subject of group theory has blossomed out in many different directions since it began as a study of permutations. It would be impossible to even hint at the many different aspects of the subject today – not only because there are so many such aspects, but also because it would require much more group theory than it has been possible to present in this book to understand merely what these aspects are.

There is one small part of the subject, however, where the facts can be easily understood even if the proofs are not. In the remainder of this chapter we will state without proof, some interesting facts about sizes of subgroups of a group.

S_4 has subgroups of size 24, 12, 8, 6, 4, 3, 2 and 1. Can you notice the special relationship that these numbers have with 24, the size of S_4? Yes, all of these numbers divide 24 exactly. This is no coincidence.

The size of a subgroup must divide the size of the group in which it lies.

This very important fact is known as Lagrange's theorem. A group with 28 elements cannot have a subgroup with 12 elements because 12 does not divide 28. The only possible sizes for a subgroup of a group with 28 elements are 1, 2, 4, 7, 14, 28, the divisors of 28.

A group with 29 elements has very few subgroups indeed, because the only divisors of 29 are 1 and 29. Therefore the only sizes of subgroups of such a group are 1 and 29. The subgroup with one element must be the subgroup consisting just of the identity and the subgroup with 29 elements must be the whole group. These are therefore the only subgroups. Moreover, if x is any element other than the identity, the cyclic subgroup generated by x must be the whole group. The group is therefore cyclic and every element, except the identity, is a generator.

The same argument can be used for any number bigger than 1 whose only divisors are 1 and itself. These numbers, 2,3,5,7,11 and so on, are called *prime numbers*. They play an important part in the theory of finite groups. For example, we have just seen that

if the size of a group is a prime number, the group is cyclic.

Similar interesting facts are that

if the size of a group is the square of a prime number, the group is Abelian

if the size of a group is a power of a prime number or the product of powers of two primes or any odd number whatsoever or twice any odd number, the group is soluble.

As this list of facts shows, a lot of information can often be discovered about a finite group simply by knowing the number of elements it contains. The size of V_4 is $4 = 2^2$, which is the square of a prime number. Therefore V_4 is Abelian (a fact that we already knew). The size of S_4 is $24 = 8 \times 3 = 2^3 \times 3^1$, which is the product of powers of two primes. Thus S_4 is soluble.

The fact that a group of odd size is soluble is a very deep theorem in the subject. It had been suspected for many years, but it was only in 1963 that it was shown to be true by a 250-page proof.

Lagrange's theorem tells us that the size of the subgroup divides the size of the group. Does this work in reverse? If a certain group has n elements and if m divides n, must the group have a subgroup of size m? The answer is 'not necessarily'. For example, although 6 divides 12 and A_4 has 12 elements, it has no subgroup with exactly 6 elements. However, although the converse of Lagrange's theorem does not always hold, it holds in certain circumstances.

If G is a group of size n and m is a divisor of n, then G has a subgroup of size m in each of the following cases:

(i) m is the power of some prime number

or (ii) G is Abelian

or (iii) G is soluble and the greatest common divisor of m and n/m is 1.

Example 1. G is a group of size 200. What are the possible sizes of its subgroups?

Solution. With only the most elementary knowledge of groups we can conclude that G has subgroups of sizes 1 and 200 (they are $\{1\}$ and G itself). Using the first case of the above theorem we may conclude that G has subgroups of size 2, 4, 8, 5 and 25 as these are all powers of primes which divide 200. Possible sizes of subgroups are therefore 1, 2, 4, 5, 8, 25 and 200. Whether G has subgroups of sizes 10, 20, 40, 50 or 100 depends on which of the many groups of size 200 G is.

Example 2. G is a soluble group of size 300. What are the possible sizes of its subgroups?

Solution. The methods used in example 1 yield the following possibilities: 1,2,3,4,5,25,300. In addition, we know that the group is soluble. Using case (iii) of the above theorem we conclude that G has subgroups of size 12, 75 and 100. For example, 75 divides 300 and the greatest common divisor of 75 and 300/75, that is, the greatest common divisor of 75 and 4, is 1. Combining these results we have concluded that G has subgroups of size 1, 2, 3, 4, 5, 12, 25, 75, 100 and 300. In addition G might, though we cannot tell, have subgroups of size 6, 15, 20, 50, 60, 150.

Example 3. G is a group of size 105. What are the possible sizes of its subgroups?

Solution. We are not told that the group is soluble, but since its size is an odd number, it must be a soluble group according to the very deep theorem quoted above. Using the methods of example 2, we conclude that G has subgroups of size 1, 3, 5, 7, 15, 21, 35 and 105. These are in fact all of the divisors of 105.

Example 4. G is an Abelian group of size 60. What are the possible sizes of its subgroups?

Solution. By case (ii) of the above theorem, G has subgroups of all possible sizes, that is, subgroups of size 1, 2, 3, 4, 5, 6, 10, 12, 15, 20, 30 and 60.

The above theorem gave situations in which the converse to Lagrange's theorem holds. What about cases where it fails? The following theorem shows that it often fails in S_n.

If m and n are two numbers such that

(i) *the greatest common divisor of m and n!/m is 1,*

and (ii) *m is not equal to 1, 144, 1152, (n–1)! or n!, and is not a power of a prime,*

then S_n *has no subgroups of size m.*

Example 5. Which divisors of 5! are impossible sizes for subgroups of S_5?

Solution. $5! = 120 = 2^3 \times 3 \times 5$. Take $n = 5$ and $m = 2^3 \times 5 = 40$. Then $n!/m = 3$, and the greatest common divisor of 40 and 3 is 1. Moreover 40 is not a power of a prime and is not equal to any of the values listed in condition (ii) above. Hence S_5 has no subgroups of size 40. Similarly we can show that S_5 has no subgroups of size $3 \times 5 = 15$.

Example 6. Which divisors of 7! are impossible sizes for subgroups of S_7?

Solution. $7! = 5040 = 2^4 \times 3^2 \times 5 \times 7$. The following values of m satisfy conditions (i) and (ii).

$$m = 5 \times 7 = 35$$
$$m = 3^2 \times 5 = 45$$
$$m = 3^2 \times 7 = 63$$
$$m = 3^2 \times 5 \times 7 = 315$$
$$m = 2^4 \times 5 = 80$$
$$m = 2^4 \times 7 = 112$$
$$m = 2^4 \times 5 \times 7 = 560$$
$$m = 2^4 \times 3^2 \times 7 = 1008$$

Hence S_7 has no subgroups with these sizes.

Note that $m = 2^4 \times 3^2 = 144$, and $m = 2^4 \times 3^2 \times 5 = 720 = 6!$ satisfy condition (i) but not condition (ii). In fact S_7 does have subgroups of sizes 144 and 720 (S_6 is an example of the latter) and so condition (ii) is really necessary!

Has S_7 a subgroup of size 105? This is not one of the values given by the above theorem. But nevertheless we can show that there can be no such subgroup. For if there was a subgroup of size 105, this in turn must have a subgroup of size 35 (see example 3). This group of size 35 will therefore be a subgroup of S_7. However 35 has been ruled out as a possibility in the table above and so 105 must be ruled out also. Thus S_7 has no subgroups of size 35, 45, 63, 80, 105, 112, 315, 560, 1008.

EXERCISE SET 16

1. Which of the following multiplication tables can be the multiplication tables for a group?

(a)

	a	b	c
a	a	b	c
b	b	c	d
c	c	d	b

(b)

	a	b	c
a	a	a	a
b	a	b	b
c	a	b	a

(c)

	a	b	c
a	c	a	b
b	a	b	c
c	b	c	a

(d)

	a	b	c
a	a	b	c
b	b	a	a
c	c	a	a

2. (a) 'A group of order 693 must have a subgroup of order n.' Find as many numbers as you can which make this statement true.
   (b) Repeat for groups of order 72.

3. What is the commutator subgroup of each of the following groups?
   (a) S_3
   (b) the group of quaternions on page 68
   (c) the group of rotations of a square on page 59

4. G is a non-soluble group and the number of elements in G is either

   142, 160, 168, 195 or 243.

   Which of these five values must it be?

5. What is the smallest non-Abelian group?

6. Which divisors of 6! *cannot* be the size of a subgroup of S_6?

7. (a) Which of the odd divisors of $10! = 2^8 \times 3^4 \times 5^2 \times 7$ are sizes of subgroups of S_{10}?
   (b) Which of the odd divisors of 10! cannot be the sizes of subgroups of S_{10}?

Solutions to the Exercises

EXERCISE SET 1

1. (a) (13524), (b) (1425), (c) (124)(35), (d) (25), (e) I, (f) (12)(45).
2. (b) is not since both $1 \to 1$ and $4 \to 1$.
3. (b) (it is the inverse of the other two).

EXERCISE SET 2

1. (a) should be (19)(2875)(364).
   (c) should be (17925)(3648).
2. I, (12), (13), (23), (123), (132).

EXERCISE SET 3

1. (a) $(xx)(xx)(xx)(xxx)$, (b) $(xx)(xx)(xxxx)$, (c) I, (d) $(xx)(xxx)(xxxx)$.
2. (a) (12), (13), (14), (15), (23), (24), (25), (34), (35), (45).
   (b) (12)(34), (12)(35), (12)(45), (13)(24), (13)(25), (13)(45), (14)(23), (14)(25), (14)(35), (15)(23), (15)(24), (15)(34), (23)(45), (24)(35), (25)(34).
   (c) (12)(345), (12)(354), (13)(245), (13)(254), (14)(235), (14)(253), (15)(234), (15)(243), (145)(23), (154)(23), (135)(24), (153)(24), (134)(25), (143)(25), (125)(34), (152)(34), (124)(35), (142)(35), (123)(45), (132)(45).
3. I, (xx), (xxx), $(xxxx)$, $(xxxxx)$, $(xxxxxx)$, $(xx)(xx)$, $(xx)(xxx)$, $(xx)(xxxx)$, $(xxx)(xxx)$, $(xx)(xx)(xx)$.

EXERCISE SET 4

1. (a) (23), (b) (124), (c) (134), (d) (1253), (e) (12)(34).
2. (a) (123), (b) (132).
3. (1324).
4.

	I	(123)	(132)	(12)	(13)	(23)
I	I	(123)	(132)	(12)	(13)	(23)
(123)	(123)	(132)	I	(23)	(12)	(13)
(132)	(132)	I	(123)	(13)	(23)	(12)
(12)	(12)	(13)	(23)	I	(123)	(132)
(13)	(13)	(23)	(12)	(132)	I	(123)
(23)	(23)	(12)	(13)	(123)	(132)	I

EXERCISE SET 5

1. (a) (145)(236), (b) (16427)(35), (c) I, (d) (12)(34)(56), (e) (1342), (f) (1243).
2. (a) (125364), (b) (146352), (c) (235)(46), (d) (162), (e) (164235), (f) (146352).
3. (a) true, (b) false, (c) true, (d) true.

EXERCISE SET 6

1. (a) is meaningless since permutations cannot be added or subtracted,
   (b), (c) and (f) can be false, e.g. $x = (123)$, $y = (12)$,
   (d) and (e) are always true.

2. $x^1 = (1342)(567)$, $x^2 = (14)(23)(576)$, $x^3 = (1243)$, $x^4 = (567)$, $x^5 = (1342)(576)$, $x^6 = (14)(23)$, $x^7 = (1243)(567)$, $x^8 = (576)$, $x^9 = (1342)$, $x^{10} = (14)(23)(567)$, $x^{11} = (1243)(576)$, $x^{12} = I$, $x^{13} = (1342)(567) = x^1$. The order of x is 12.

3. (a) 5, (b) 10, (c) 6, (d) 2, (e) 1.

4. (12), (13), (14), (23), (24), (34), (12)(34), (13)(24), (14)(23).

5. (a) 2, (b) 6, (c) 4, (d) 3, (e) 5.

6. (a) $(x^2)^2 = x^4$, $(x^2)^3 = x^6$, $(x^2)^4 = x^8 = I$.
(b) $(x^6)^2 = x^{12} = x^4$, $(x^6)^3 = x^{18} = x^2$, $(x^6)^4 = x^{24} = I$.
(c) $(x^5)^2 = x^{10} = x^2$, $(x^5)^3 = x^{15} = x^7$, $(x^5)^4 = x^{20} = x^4$, $(x^5)^5 = x^{25} = x$, $(x^5)^6 = x^{30} = x^6$, $(x^5)^7 = x^{35} = x^3$, $(x^5)^8 = x^{40} = I$.

EXERCISE SET 7

1. 12. N.B. $w = (ma)^4 m(ma) = (1345)(267)$.

2. (a) 6 weeks later, (b) 6 weeks later.
N.B. Now, $m = (123457)$, $a = (1357)(24)$

EXERCISE SET 8

1. $f = (124875)(36)$.

2. $f^2 = (147)(285) = l$, $d = r^4 = (lc)^4 = (f^2c)^4$.

EXERCISE SET 9

1. (a) (13), (b) (12)(346), (c) (15)(34), (d) (2435), (e) (156).

2. (a) (124), (c) (23), (e) (23) (other answers are possible). In (b) and (d), x and z are not conjugate.

3. (a) (123), (b) (1234), (c) (1234567).

EXERCISE SET 10

(In all of the exercises of this set, there are other correct answers though the ones given are the ones obtained by following the methods suggested in the text.)

1. (15)(12)(16)(37)(34).

2. $(AB^4)^5B^3AB^2$ which can be simplified further to BAB^3AB^2 since AB^4 has order 6.

3. $A(CA)^5A(CA)^4A(CA)^7A(CA)^7A(CA)^4A(CA)^5$.

4. $B^2F^3B^3F^3BF^3B^3F^3BF^3$.

EXERCISE SET 11

(In all of the exercises of this set, there are other correct answers.)

1. (a) $l = (165432)$, $t = (14)(23)$, $(tl)^3 = (13)$, $tl^3 = (2635)$, $(16) = (tl)^3$ conjugated by $(tl^3)^{-1} = tl^3tltltl^4t$ after simplification.
(b) Impossible.
(c) Impossible.
(d) $l = (1765432)$, $t = (13)$, $tl = (12)(37654)$, $(tl)^5 = (12)$, $tl^2 = (275364)$, $(17) = (tl)^5$ conjugated by $tl^2 = l^6tltltltl^2$ after simplification.

2. For convenience put 1 = D, 2 = E, 3 = A, 4 = R, 5 = G, 6 = R, 7 = A, 8 = N. Desired permutation $= (185627)(34) = (tl)^2 tl^3 (tl)^5 l^5 t(l^7 t)^2 (l^7 t)^2 l^5 t(tl)^5 tl^3 (tl)^2 tl^3 (tl)^5 (l^5 t)^2 (l^7 t)^2 l^5 t(tl)^5 tl^3 (tl)^7 tl^3 (tl)^5 (l^5 t)^2 (tl)^5 (tl^3)^2 (tl)^5$.

3. Basic operations:
r = move each coin clockwise one space around perimeter of hole = (1234567)
j = jump coin over hole from right to left = (1234).
N.B. After each move bring empty space back to left-hand end of middle row by moving some of the coins clockwise.
$rj = (13)(24567)$, $(rj)^5 = (13)$, $r^2 j^2 = (35746)$, $(rj)^2 = (25746)$. Desired permutation = (17)(26)(35) $= (j^2 r^5)^2 (rj)^5 (r^2 j^2)^2 (j^3 r^5)^2 r^2 j^2 (rj)^5 j^2 r^5 (rj)^2 r^2 j^2 (rj)^5 (j^2 r^5)^2 (j^3 r^5)^2 r^2 j^2 (rj)^5 j^2 r^5 (rj)^7 j^2 r^5 (rj)^5 r^2 j^2 (rj)^5$.

4. First operation can be summarized as: 6→ right-hand end, 1→6, 4→1, 7→4, 2→7, right-hand end →2 = a = (16274).
N.B. These numbers refer to positions and not to the particular pieces.
Second operation can be summarized as: 6→ right-hand end, 3→6, 8→3, 5→8, 2→5, right-hand end →2 = b = (25836). The third operation is c = (26) which can be achieved by 6→ right-hand end, 2→ left-hand end, right-hand end →2, left-hand end →6. Desired permutation = (17)(26)(35)(48) $= aca^2 ca^3 ca^4 cbcb^2 cb^3 cb^4 a^2 ca^3 b^2 cb^3 a^2 ca^3$.

EXERCISE SET 12

1. (a), (c), (e).
2. (c), (d), (e).
3. $P[x^{-1}y^{-1}xy] = P[x^{-1}]P[y^{-1}]P[x]P[y] = P[x]P[y]P[x]P[y] = P[x]^2 P[y]^2 = 1$.
4. Case (b) is impossible because l = (15432), t = (14)(23) both of which are even. However the desired permutation, (15), is odd. To show that case (c) is impossible requires a knowledge of 'package-preserving permutations' as discussed in the next chapter.

EXERCISE SET 13

1. (c), (e) are groups. (a) and (d) are not groups because they do not contain I, and (b) is not a group because it is not closed. For example, $(1234)^2 = (13)(24)$ which is not in the group.
2. A_4.

EXERCISE SET 14

1. (b), (d).
2. No! Although it contains 5 permutations including some of order 5, it is not a cyclic group because it is not a group at all. For example, (15432)(13254) = (14235) which is not in the set.
3. (a) {I, (134), (143)}; (b) {I, (134)(56), (143), (56), (134), (143)(56)}.
4. {I, (12), (13), (23), (132), (123), (45), (12)(45), (13)(45), (23)(45), (132)(45), (123)(45)}.
5. I, (12)(34), (12)(56), (34)(56), (13)(24), (14)(23), (15)(26), (16)(25), (35)(46), (36)(45), (135)(246), (136)(245), (145)(236), (146)(235), (153)(264), (154)(263), (163)(254), (164)(253), (1324)(56), (1423)(56), (1526)(34), (1625)(34), (12)(3546), (12)(3645).

EXERCISE SET 15

1. S_3.
2. {I, (13), (24), (13)(24)}.

3. (a) 10, (b) 4, (c) 8, (d) 2.

4. $\{$I, (14)(36), (1245), (14)(25), (1542), (12)(36)(45), (25)(36), (15)(24)(36)$\}$, 90.

5. 60, 7983360.

EXERCISE SET 16

1. (c) is the only group table. In (a), *a* is the identity and *b* and *c* have no inverses. In (b) there is no identity. In (d), *a* is the identity but *b* and *c* have two inverses each, impossible for a group.

2. (a) 1, 3, 7, 9, 11, 63, 77, 99, 693.
   (b) 1, 2, 3, 4, 8, 9, 72.

3. (a) A_3, (b) $\{1,-1\}$, (c) $\{$I, (13)(24)$\}$.

4. 168. (195 and 243 are odd and 142 and 160 are products of two powers of prime numbers.)

5. S_3.

6. The impossible values that can be deduced from the theorem are 45 and 80 (though there are others).

7. (a) 1, 3, 5, 7, 9, 25, 27, 81,
   (b) 175, 525, 567, 1575, 2025, 2835, 4725, 14175.